LONDON MATHEMATICAL SOCIETY STUDENT TEXTS

Managing Editor: C. M. SERIES, Mathematics Institute, University of Warwick
Coventry CV4 7AL, United Kingdom

London Mathematical Society Student Texts 38

Harmonic Maps, Loop Groups, and Integrable Systems

Martin A. Guest
University of Rochester

CAMBRIDGE
UNIVERSITY PRESS

CAMBRIDGE UNIVERSITY PRESS
Cambridge, New York, Melbourne, Madrid, Cape Town, Singapore, São Paulo

Cambridge University Press
The Edinburgh Building, Cambridge CB2 8RU, UK

Published in the United States of America by Cambridge University Press, New York

www.cambridge.org
Information on this title: www.cambridge.org/9780521580854

First published 1997

A catalogue record for this publication is available from the British Library

Library of Congress Cataloguing in Publication data

Guest, Martin A.

Harmonic maps, loop groups, and integrable systems / Martin A.
Guest.

p.   cm. – (London Mathematical Society student texts : 38)

Includes bibliographical references (p.   –   ) and index:

ISBN 0-521-58085-4. (hardback). – ISBN 0-521-58932-0 (paperback)

1. Harmonic maps.   2. Loops (Group theory)   3. Differential
equations.   I. Title.   II. Series.
QA614.73.G84   1997   96-38837
514′.74 – dc20   CIP

ISBN 978-0-521-58085-4 hardback
ISBN 978-0-521-58932-1 paperback

Transferred to digital printing 2007

Contents

Part III
One-dimensional and two-dimensional integrable systems

Preface

Motivation: Harmonic maps

The principal motivation for this book was provided by certain recent advances in the theory of harmonic maps, which depend on ideas from the theory of integrable systems.

The concept of harmonic map is a generalization of the concept of geodesic. The harmonic maps in this book go from Riemann surfaces to compact Lie groups or symmetric spaces; they are, therefore, two-dimensional analogues of geodesics. They encompass many fundamental examples in differential geometry, such as minimal surfaces, which have been studied by geometers over a long period of time. Nevertheless, important new discoveries are still being made, and major open problems remain. Since the late 1970s, the field has acquired new vitality from mathematical physics, in the guise of the non-linear sigma model or chiral model. As a result, harmonic maps have attracted the attention of a much wider audience than before, both within the mathematical community and beyond.

One of the themes of research in this area during the last 25 years or so is the "classification" of such harmonic maps, i.e., the description (or parametrization) of harmonic maps from Riemann surfaces to compact Lie groups or symmetric spaces, in terms of well known maps. There are several reasons for doing this. The most obvious one is that such a description provides the "general solution" of the relevant harmonic map equation. Another is that such a description should be useful in describing the "moduli space" (or, more accurately, the parameter space) of solutions of the harmonic map equation. The nature of this problem turns out to be algebraic, rather than analytical; it is global rather than local. The methods used to study the problem are therefore closer to algebra and topology than to analysis. There are of course other problems of interest, and progress has been made with these as well. But this book is concerned entirely with the classification problem. The survey articles of Eells and Lemaire [1978; 1988] may be consulted for a modern view of the history of harmonic maps and its problems, including the progress made on the classification problem before 1988.

A turning point in the theory was the idea that the harmonic map equation is a kind of "integrable system". This idea first arose explicitly in the mathematical physics literature, for example in Pohlmeyer [1976]; Zakharov and Mikhailov [1978]; Zakharov and Shabat [1979]. The harmonic map equation was reformulated as a kind of Lax equation "with parameter". This was taken up in Uhlenbeck [1989], where the first new results were obtained. At approximately the same time, great progress was being made with integrable systems such as the KdV equation by employing certain infinite dimensional Lie algebras (affine Kac-Moody Lie algebras) and Lie groups (loop groups). The potential of these Lie theoretic methods in the harmonic map problem was suggested in Segal [1989], where another proof of Uhlenbeck's results was given.

Further developments came rather slowly, reflecting the fact that the harmonic map problem is quite different from those problems like the KdV equation which had been tackled so successfully. Lax equations and loop groups provided a solution of the classification problem for harmonic maps from S^2 to the unitary group U_n (in Uhlenbeck [1989]; Valli [1988]; Wood [1989]; Segal [1989]), where "solution" is taken to mean "reduction to holomorphic data". However, the question of generalization – to higher genus Riemann surfaces, or other Lie groups and symmetric spaces – still seemed inaccessible, and in any case the solution for U_n did not amount to giving explicit formulae for harmonic maps. Thus, the "integrable systems approach" seemed at this point not to provide the breakthrough that had been hoped for.

A breakthrough came in the genus one case, i.e., harmonic maps from a torus. Wente's counterexample to the Hopf conjecture (Wente [1986]) was shown to arise from the integrable systems point of view in Pinkall and Sterling [1989]. This was generalized significantly by various authors (for example, in Burstall *et al.* [1993]), thus providing further evidence of the usefulness of integrable systems and loop groups. Meanwhile, the loop group approach to the genus zero case (i.e., harmonic maps from S^2) was pursued in Bergvelt and Guest [1991]; Guest and Ohnita [1993]. This yielded new results on the connected components and fundamental group of the space of harmonic maps, in which the role of the "symmetry group" is essential.

With the benefit of hindsight, and a new point of view suggested in Dorfmeister *et al.* [to appear], the genus zero case turns out to have some aspects in common with the genus one case. This leads to new explicit formulae in the genus zero case, and also a way of generalizing the results of Uhlenbeck and Segal. It seems reasonable to conclude that the basis of a unifying theory has now been been established. From a naive computational point of view, the basic phenomenon is that the harmonic maps considered here may be expressed in terms of "factorization of exponentials".

This book is based on lectures in which I attempted to present this unifying theory in a straightforward manner. I certainly failed in this attempt, as the time available in each series of lectures permitted only about a third of the material to be covered. In the present more comprehensive exposition, however, I have tried to maintain some of the informal style of the lectures. Wherever possible, generality and completeness have been sacrificed in favour of accessibility.

Integrable systems, loop groups, and harmonic maps each have their own well developed literature, and it was not my intention merely to duplicate this. However, I hope that the modest introductions to the theories of integrable systems and loop groups given here may be useful, even for the reader whose main interest is not in harmonic maps.

A summary of the topics will be given shortly. For the benefit of the knowledgeable reader, Chapter 26 contains a more technical summary, together with information on topics which are not treated in this book.

What is an integrable system?

Whoever wishes to learn something about "integrable systems" faces at least two difficulties. One is the breadth of the subject: It ranges over mechanics, differential equations, global analysis, algebraic geometry, and Lie theory; and these are just some of the mathematical aspects, ignoring the vast intersection with physics. Moreover – this is the second difficulty – the subject accommodates a whole range of points of view, from very "pure" to very "applied". Some authors emphasize general concepts such as symplectic or Poisson structures. They naturally focus attention on those examples which best illustrate the particular general theory under discussion. Other authors give priority to the examples themselves – classical mechanical systems, the Toda lattice, the sine-Gordon equation, or the KdV equation perhaps – referring to the general theories only in passing. Needless to say, this leads to quite different styles of exposition, which can be disconcerting for the beginner.

There is one common thread in all this, namely the idea of a symmetry group of a differential equation. The nicest and most natural equations (such as those which occur in the "real world") often admit symmetry groups. The existence of a large enough symmetry group leads to the possibility of solving the differential equation by *algebraic* means, and this is perhaps the fundamental property of an integrable system.

The term "integrable system" is used rather loosely in the literature, and a precise definition will not be given in this book. But this is not really a practical disadvantage; after all, there is no precise definition of an elephant, yet in practice one may recognize an elephant by various fundamental properties. Similarly, one has various fundamental properties of an integrable system: It is a differential equation, usually with geometrical or physical significance, admitting a large symmetry group, which is solvable (integrable) by algebraic means, etc. As another substitute for a precise definition, one could exhibit a typical example, such as the Toda lattice. Whatever an integrable system is, the Toda lattice certainly is one! For this reason, the Toda lattice plays a prominent role in this book. Another reason for choosing the Toda lattice here, however, is that it happens to be *directly* related to (certain kinds of) harmonic maps.

This book deals only with that part of the theory of integrable systems which is immediately applicable to the classification of harmonic maps from Riemann surfaces to Lie groups or symmetric spaces. Even this is not treated comprehensively here; to do so would have stretched the book (and its author) too far. The biggest omission is the theory of spectral curves, and indeed the entire algebraic geometry point of view. Another major gap is the theory of harmonic maps from Lorentzian (rather than Riemannian) surfaces, which is closely related to soliton theory.

As far as the general theory of integrable systems is concerned, this book is even less ambitious. Nevertheless, the topics which arise here – such as Lax equations, zero-curvature equations, hidden symmetry groups, Bruhat decompositions, τ-functions, Riemann-Hilbert problems, dressing transformations – do illustrate some important features of the theory.

Brief description of topics

Part I One-dimensional integrable systems.

The first two chapters give a very brief introduction to Lie groups and Lie algebras, concentrating on the exponential map and the adjoint representation, in the context of matrix groups. Chapter 3 introduces a fundamental technical tool which will be used repeatedly: the Iwasawa decomposition of a Lie group.

Chapter 4 discusses an important classical source of motivation, Hamilton's equations. The concept of a general Hamiltonian system is illustrated by a famous example: a "height function" on an adjoint orbit of a compact Lie group. The Hamiltonian function, its corresponding Hamiltonian vector field, and the associated one parameter group of diffeomorphisms are easily described and explicitly computed for this example.

Chapters 5 to 8 constitute a survey of the (one-dimensional, open) Toda lattice, a "model" integrable system. This is simply a first order ordinary differential equation, but with a very interesting algebraic structure. The Lax form of the equation is given in Chapter 5, from which the general solution is obtained explicitly. The formula for this solution involves exponentiating a (linear) matrix function, then performing a certain matrix factorization.

In Chapter 6, the story is repeated, but this time entirely in Lie algebraic terms. This paves the way for a discussion of the generalized Toda lattice (for a general Lie algebra, and for general invariant Hamiltonian functions) in Chapter 7. The role of the classical Toda lattice as just one member of a hierarchy of integrable systems is thus revealed. Chapter 8 discusses further generalizations, such as the concept of an R-matrix, and a curious relationship between the Toda lattice and the example of Chapter 4.

Part II Two-dimensional integrable systems.

Chapters 9 to 22 are intended to be parallel to Chapters 1 to 8, but using infinite dimensional Lie groups instead of finite dimensional Lie groups, zero-curvature equations instead of Lax equations, and the two-dimensional Toda lattice (and harmonic map equation) instead of the one-dimensional Toda lattice.

The first two chapters provide motivation. Zero-curvature equations are treated as two-dimensional analogues of Lax equations, and our two main examples are introduced (the periodic two-dimensional Toda lattice, and the harmonic map equation for maps from surfaces to Lie groups). There is now little hope of writing down "the general solution". As a substitute, one has the method of dressing transformations, which converts a trivial solution into a non-trivial solution, by applying an element of an infinite dimensional Lie group. It is at this point that loop groups appear.

Chapter 11 gives the bare facts concerning loop groups and loop algebras which will be needed. Chapter 12 gives an equally brief description of the Iwasawa decomposition for loop groups.

The two-dimensional Toda lattice is the subject of Chapters 13 to 15. In Chapter 13 the equation is formulated precisely in loop theoretic terms. In Chapters 14 and 15 a family of explicit solutions is obtained by applying dressing transformations to a trivial solution. (Chapter 14 is devoted to an elementary explanation of τ-functions, primarily for the one-dimensional case, as preparation for Chapter 15.)

Harmonic maps from surfaces to compact Lie groups and symmetric spaces are the subject of Chapters 16 to 22. Chapter 16 gives the precise loop theoretic formulation for harmonic maps into Lie groups, and Chapter 18 does the same for symmetric spaces. In each case a suitable symmetry group is identified, which acts on solutions (i.e., harmonic maps) by dressing transformations. Chapters 17 (for Lie groups) and 19 (for symmetric spaces) describe some well known results which are available when the domain surface is S^2, or more generally when the harmonic maps have "finite uniton number".

In Chapter 20 a geometrical method for studying harmonic maps is introduced, based on the Bruhat decomposition of the Grassmannian model of a loop group. As illustrations, a new proof is given of the classification of harmonic maps from S^2 into complex projective space, together with some new estimates for the "uniton number" of harmonic maps from S^2 into complex Grassmannians.

Chapter 21 discusses primitive maps and their relevance to harmonic maps, and a relationship between the two-dimensional Toda lattice and harmonic maps.

Using the method of Chapter 20, it is shown in Chapter 22 how the harmonic maps obtained so far are given by formulae which are surprisingly analogous to the "factorization of exponentials" formulae appearing as the solutions to the one-dimensional Toda lattice. This sets the scene for Part III.

Part III One-dimensional and two-dimensional integrable systems.

Chapters 23 to 25 bring together Parts I and II by showing that they are not just analogous, but in fact very directly related. This direct relationship is based on the construction (described in Chapter 23) of a solution of a zero-curvature equation from the solutions of two Lax equations, the "$1+1=2$" principle.

In Chapter 24 this is applied to the harmonic map equation; the solutions obtained by this method are called harmonic maps of "finite type". Like all our previous examples, they are given by "factorization of exponentials" formulae. More generally, one has primitive maps of finite type.

Harmonic maps of finite type are quite different from harmonic maps of finite uniton number, and their study is only just beginning. In Chapter 25 the first important example is described, namely harmonic maps of finite type from a torus to S^2.

Background knowledge required

Only the following knowledge will be taken for granted: linear algebra, elementary definitions of topology, basic theory of ordinary differential equations, and elementary properties of differentiable manifolds (such as the concepts of tangent bundle and vector field).

I have tried not to assume too much in the way of Lie theory, or the theory of loop groups, or for that matter the theory of harmonic maps; but the reader who has some knowledge of at least one of these areas will find the material much easier to read.

There exist beautiful generalizations of many aspects of the theory presented here, but I have resisted the temptation to pursue them. For example, most of the time I consider only matrix groups, primarily the orthogonal and unitary groups, rather than general Lie groups. I have left symplectic and Poisson geometry in the background, where they certainly belong in an introductory course.

The exercises are quite numerous in Part I, where it is hoped that they will be more useful. They die out rapidly thereafter. On the other hand, proofs become progressively more detailed. It is an attractive feature of the topics discussed here that the proofs are usually quite elementary and devoid of mysterious technicalities. In the initial chapters, however, most proofs are of standard results, and so they are sketched very briefly or omitted.

The references include most of the recent literature on *connections* between harmonic maps, loop groups, and integrable systems. But within each of these three areas, the references are not at all comprehensive. In many cases I have referred to books rather than original articles; I found Perelomov [1990] and Fordy and Wood [1994] particularly helpful. Other important books that should be mentioned are Carter *et al.* [1995] (for an accessible introduction to Lie theory); Kac [1990]; Pressley and Segal [1986] (for infinite dimensional Lie algebras and Lie groups); Urakawa [1993] (for harmonic maps); and Arnold [1978]; Arnold and Novikov [1990; 1994]; Faddeev and Takhtajan [1987]; Fomenko and Trofimov [1988]; Gu [1995]; Guillemin and Sternberg [1984]; Newell [1985]; Novikov *et al.* [1984] (for integrable systems). The bibliographical comments at the ends of appropriate chapters provide some pointers to the original sources. I apologize for the inevitable omissions here, particularly of the Russian and Japanese literature.

Acknowledgements

My attempts to understand some of the relationships between integrable systems, loop groups, and harmonic maps would not have progressed very far without the help of Maarten Bergvelt, Francis Burstall, Yoshihiro Ohnita, and Andrew Pressley. I am greatly indebted to them for their collaboration.

While preparing the first version of these notes in Japan, I was able to discuss many aspects of integrable systems with Reiko Miyaoka and Takashi Otofuji. I am very grateful to them for this opportunity, and indeed to all members of the Department of Mathematics at Tokyo Institute of Technology for their hospitality.

I am very grateful to Yoshiaki Maeda for inviting me to give lectures at Keio University, and to Mutsuo Oka for inviting me to give lectures at Tokyo Institute of Technology, in 1994. I thank the audiences of both series of lectures for their participation, comments, and questions. My stay in Tokyo was financed in part by a grant from the U.S. National Science Foundation, under the NSF-CGP Program.

John Bolton, Mike Gage, William Liu, Amos Ong, Franz Pedit, John Wood, and Lyndon Woodward made helpful comments and suggestions which were greatly appreciated. In addition to the books and articles listed in the references, I benefitted from unpublished lecture notes of N.M. Ercolani and H. Flaschka. Work on this project was completed with the aid of a research grant from the NSF.

This book was typeset using $\mathcal{AMS}$-TEX.

Martin Guest

Rochester

Part I
One-dimensional integrable systems

Chapter 1: Lie Groups

I. Definitions and examples.

We assume that the reader is familiar with the idea of a (smooth) manifold. For our purposes, it will be enough to think of a manifold which is embedded (as a submanifold) in some $\mathbf{R}^n$. For example, the sphere $S^{n-1} = \{x \in \mathbf{R}^n \mid ||x|| = 1\}$ is embedded in $\mathbf{R}^n$.

We assume also that the reader understands the concept of tangent bundle. If X is a submanifold of $\mathbf{R}^n$, the tangent space to X at a point $x \in X$ is the space of "tangent vectors to curves in X through x", i.e.,

$$T_x X = \{\gamma'(0) \mid \gamma : (-\epsilon, \epsilon) \to \mathbf{R}^n, \gamma(-\epsilon, \epsilon) \subseteq X, \gamma(0) = x\}.$$

Here, $\gamma'(0) = \frac{d}{dt}\gamma(t)|_0 = D\gamma_0(\frac{d}{dt})$. (In the case of S^{n-1}, it is easy to verify the usual description of $T_x S^{n-1}$ as the set of all vectors which are orthogonal to x.) The tangent bundle of X can then be defined as

$$TX = \{(x, U) \in X \times \mathbf{R}^n \mid U \in T_x X\}.$$

This is a submanifold of $X \times \mathbf{R}^n$, and there is a natural projection map $\pi : TX \to X$. If $f : X \to Y$ is a (smooth) map, then the derivative of f is a map $Df : TX \to TY$.

Using the tangent bundle, one can define another basic object: A vector field on X is a (smooth) map $V : X \to TX$ such that $\pi \circ V$ is the identity map of X. The value of V at x will be denoted by V_x; V_x is an element of $T_x X$. An important property of vector fields is that they act on (smooth) functions "by differentiation": If V is a vector field, and $f : X \to \mathbf{R}$, then we obtain a function Vf by means of the formula $(Vf)(x) = (Df)_x(V_x)$. Here we use the standard convention that $T_u \mathbf{R}$ is identified canonically with $\mathbf{R}$, for any $u \in \mathbf{R}$.

Definition. *Let X be a (smooth) manifold. We say that X is a (real) Lie group if*

(1) X has a group structure, $\circ$, and

(2) the map $X \times X \to X$, $(x, y) \mapsto x \circ y^{-1}$ is smooth.

A complex Lie group may be defined in a similar way: It is a complex manifold X, with a group structure, such that the map $X \times X \to X$, $(x, y) \mapsto x \circ y^{-1}$ is complex analytic (holomorphic).

The main examples of Lie groups are *matrix groups*.

Examples:

(1) $M_n\mathbf{R} = \{$real $n \times n$ matrices$\}$, $\circ$ = matrix addition. Similarly for $M_n\mathbf{C}, M_n\mathbf{H}$. More generally, any real (or complex) vector space is a real (or complex) Lie group.

(2) $GL_n\mathbf{R} = \{A \in M_n\mathbf{R} \mid A$ is invertible$\}$, $\circ$ = matrix multiplication. Similarly for $GL_n\mathbf{C}$.

(3) $SL_n\mathbf{R} = \{A \in GL_n\mathbf{R} \mid \det A = 1\}$, $\circ$ = matrix multiplication. Similarly for $SL_n\mathbf{C}$.

(4) $O_n = \{A \in M_n\mathbf{R} \mid A^t = A^{-1}\}$, $\circ$ = matrix multiplication. Similarly we have $U_n = \{A \in M_n\mathbf{C} \mid A^* = A^{-1}\}$, $Sp_n = \{A \in M_n\mathbf{H} \mid A^* = A^{-1}\}$.

(5) $SO_n = \{A \in O_n \mid \det A = 1\}$, $\circ$ = matrix multiplication. Similarly for SU_n.

Exercises:

(1.1) In the above list of examples, which are real Lie groups? Which are complex Lie groups?

(1.2) Why are $GL_n\mathbf{H}, SL_n\mathbf{H}$ and SSp_n omitted from the above list of examples?

(1.3) Show that $GL_n\mathbf{C}$ is connected.

(1.4) Show that SO_n is compact and connected.

(1.5) Show that O_n has two connected components, each of which is diffeomorphic to SO_n. Is it true that O_n is isomorphic (as a group) to $SO_n \times \{\pm I\}$?

Definition. *Let G_1, G_2 be Lie groups. Let $\Theta : G_1 \to G_2$ be a (smooth) map. We say that Θ is a (Lie group) homomorphism if $\Theta(gh) = \Theta(g)\Theta(h)$ for all $g, h \in G$.*

Similarly, we define the concepts of monomorphism, epimorphism, isomorphism, etc.

Example:

The determinant map det : $GL_n\mathbf{R} \to \mathbf{R}^*$ is a homomorphism. (Here, $\mathbf{R}^* = GL_1\mathbf{R}$, the group of non-zero real numbers.)

The concept of "subgroup" requires a little care:

Definition. *Let G_1, G_2 be Lie groups, such that G_1 is an (algebraic) subgroup of G_2. We say that G_1 is a Lie subgroup of G_2 if the inclusion map $G_1 \to G_2$ is an embedding.*

If G_1 is an (algebraic) subgroup of G_2, and also a submanifold, then G_1

is certainly a Lie subgroup of G_2. However, for reasons which will become clear in the next chapter, we do not *insist* that a Lie subgroup should also be a submanifold. The standard example of this is given by the (algebraic) subgroup $G_1 = \{[ta, tb] \mid t \in \mathbf{R}\}$ of the Lie group $G_2 = \mathbf{R}^2/\mathbf{Z}^2$ (for a fixed choice of $a, b \in \mathbf{R}$ with $(a, b) \neq (0, 0)$). We give G_1 the structure of a Lie group by using the natural homomorphism $\mathbf{R} \to G_1$, $t \mapsto [ta, tb]$. There are two cases to consider: (1) G_1 is isomorphic (as a Lie group) to $\mathbf{R}/\mathbf{Z}$ if $b = 0$ or if a/b is rational; (2) G_1 is isomorphic to $\mathbf{R}$ if a/b is irrational. In both cases, G_1 is a Lie subgroup of G_2. But only in case (1) is G_1 a submanifold of G_2.

It is well known that a Lie subgroup is a submanifold if and only if it is closed (see Varadarajan [1984], Theorem 2.5.4). From Chapter 3 onwards, we shall usually abbreviate the expression "closed Lie subgroup" to "subgroup".

II. The exponential map.

Let G be a Lie group. We use the following standard notation:

e = the identity element of G (= I if G is a matrix group)

$\mathbf{g} = T_e G$ = the tangent space to G at e.

The relationship between G and $\mathbf{g}$ is very important. It is useful, therefore, to have an explicit description of $\mathbf{g}$. Here are some examples, for matrix groups:

Examples:

(1) $T_e M_n \mathbf{R} = M_n \mathbf{R}$ (because $M_n \mathbf{R}$ is a vector space).

(2) $T_e GL_n \mathbf{R} = M_n \mathbf{R}$ (because $GL_n \mathbf{R}$ is an open subset of a vector space).

(3) $T_e O_n = \text{skew}_n \mathbf{R} = \{A \in M_n \mathbf{R} \mid A^t = -A\}$.

(Proof: If $X \subseteq \mathbf{R}^n$, we use the description of $T_x X$ given earlier, i.e., the space of tangent vectors $\gamma'(0)$ with $\gamma : (-\epsilon, \epsilon) \to X \subseteq \mathbf{R}^n$ and $\gamma(0) = x$. In the case of $O_n \subseteq M_n \mathbf{R}$, $x = I$, we have $\gamma(t)^t \gamma(t) = I$ for all $t \in (-\epsilon, \epsilon)$. By differentiation, we obtain $\gamma'(0)^t \gamma(0) + \gamma(0)^t \gamma'(0) = 0$, hence $\gamma'(0)^t = -\gamma'(0)$. Thus, $T_e O_n \subseteq \text{skew}_n \mathbf{R}$. Conversely, if $A \in \text{skew}_n \mathbf{R}$, let $\gamma(t) = \exp tA$. Then we have $\gamma : \mathbf{R} \to M_n \mathbf{R}$, such that $\gamma(\mathbf{R}) \subseteq O_n$ and $\gamma(0) = I$. By differentiation, $\gamma'(0) = A$. Hence $\text{skew}_n \mathbf{R} \subseteq T_e O_n$.)

The (matrix) exponential function

$$\exp A = \sum_{n \geq 0} \frac{A^n}{n!},$$

which appeared in Example (3), is very useful. (It is easy to show that the series converges for any A.) It has the following properties:

Proposition.

(1) The exponential map $\exp : M_n\mathbf{R} \rightarrow GL_n\mathbf{R}$ *is a local chart at* $I \in GL_n\mathbf{R}$.

(2) Let G be a Lie subgroup of $GL_n\mathbf{R}$. Then the exponential map restricts to a map $\mathbf{g} \rightarrow G$, and this map is a local chart at $I \in G$. ∎

(This proposition may be proved by calculating the derivative of the exponential map at $0 \in M_n\mathbf{R}$. If f is any (smooth) function, then the derivative Df is given by the formula $(Df)_x(V) = \frac{d}{dt}f(\gamma(t))|_0$, where $V = \frac{d}{dt}\gamma(t)|_0$. Hence, $(D\exp)_0 A = \frac{d}{dt}\exp tA|_0 = A$. Thus, $(D\exp)_0$ is the identity map!)

More generally, it is possible to define $\exp : \mathbf{g} \rightarrow G$ for *any* Lie group. We shall not need the general definition. However, the following useful property of the exponential map (which is valid also in the general case) should be noted: For any $X \in \mathbf{g}$, the map $\gamma : t \mapsto \exp tX$ provides a curve in G with $\gamma(0) = e$ and $\gamma'(0) = X$.

Exercises:

(1.6) Let $\gamma, \delta : (-\epsilon, \epsilon) \rightarrow M_n\mathbf{R}$. Show that $(\gamma + \delta)' = \gamma' + \delta'$ and $(\gamma\delta)' = \gamma'\delta + \gamma\delta'$.

(1.7) Let $A, B \in M_n\mathbf{R}$. Is it true that $\exp A \exp B = \exp(A + B)$?

(1.8) Show that $T_e U_n = \text{skewHerm}_n \mathbf{C} = \{A \in M_n\mathbf{C} \mid A^* = -A\}$.

(1.9) Show that $T_e SL_n\mathbf{R} = \{A \in M_n\mathbf{R} \mid \text{trace } A = 0\}$.

(1.10) Show that the exponential map $\exp : \text{skewHerm}_n \rightarrow U_n$ is surjective.

Bibliographical comments.

See the comments at the end of Chapter 3.

Chapter 2: Lie Algebras

I. Definitions and examples.

Definition. *Let V be a vector space (real or complex). We say that V is a Lie algebra (real or complex) if*

(1) there is a bilinear map $[\ ,\ ] : V \times V \to V$, such that

(2) $[X, Y] = -[Y, X]$ and $[[X, Y], Z] + [[Y, Z], X] + [[Z, X], Y] = 0$ for all $X, Y, Z \in V$.

Examples:

(1) $V =$ any vector space, $[X, Y] = 0$ for all $X, Y \in V$. (In this case, we say that V is *abelian*.)

(2) $V =$ the vector space consisting of all vector fields on a manifold M, $[\ ,\ ] =$ bracket of vector fields. (Recall that the bracket of vector fields V_1, V_2 on any manifold M is defined by $[V_1, V_2]f = V_1(V_2 f) - V_2(V_1 f)$, where $f : M \to \mathbf{R}$ is any function.)

(3) $V = \mathbf{R}^3$, $[X, Y] = X \times Y$ (vector cross product).

(4) $V = \mathrm{End}(W)$, the vector space of all endomorphisms of a vector space W, $[X, Y] = XY - YX$. (If $V = \mathbf{R}^n$ or $\mathbf{C}^n$, $\mathrm{End}(V) = M_n\mathbf{R}$ or $M_n\mathbf{C}$.)

Let G be a Lie group. Let $\mathbf{g} = T_e G$. We can construct a Lie algebra structure for $\mathbf{g}$, as follows. For any $X \in \mathbf{g}$, we define a vector field X^* on G by

$$X_g^* = DL_g(X) = \tfrac{d}{dt} g \exp tX|_0.$$

(Here $L_g : G \to G$ is given by $L_g(h) = gh$. Later, we shall need the analogous map $R_g : G \to G$, $R_g(h) = hg$.) In this way, we can identify $\mathbf{g}$ with a subspace of the Lie algebra of all vector fields on G (see Example (2)). This subspace consists precisely of the *left-invariant* vector fields on G, i.e., the vector fields V such that $V_{gh} = DL_g(V_h)$ for all $g, h \in G$.

Lemma. *If U, V are left-invariant vector fields on G, then so is $[U, V]$.*

Proof. For any $f : G \to \mathbf{R}$ we have

$$V_{gh}f = Vf(gh) = Vf \circ L_g(h)$$
$$DL_g(V_h)f = DL_g(V)f(h) = Df(DL_g(V))(h) = V(f \circ L_g)(h),$$

so the condition $V_{gh} = DL_g(V_h)$ (for all $h \in G$) is equivalent to the condition $Vf \circ L_g(h) = V(f \circ L_g)(h)$ (for all $h \in G$, $f : G \to \mathbf{R}$). If

U, V satisfy this condition, we have

$$
\begin{aligned}
[U,V](f \circ L_g) &= U(V(f \circ L_g)) - V(U(f \circ L_g)) \\
&= U((Vf) \circ L_g) - V((Uf) \circ L_g) \\
&= (U(Vf)) \circ L_g - (V(Uf)) \circ L_g \\
&= ([U,V]f) \circ L_g.
\end{aligned}
$$

So $[U,V]$ satisfies the same condition. ∎

It follows that **g** inherits the structure of a Lie algebra; it becomes a subalgebra of the Lie algebra of all vector fields. We therefore obtain $[\ ,\ ] : \mathbf{g} \times \mathbf{g} \to \mathbf{g}$ (satisfying conditions (1) and (2) above). By definition we have $[X^*, Y^*] = [X,Y]^*$, for any $X, Y \in \mathbf{g}$. In future, we call **g** the *Lie algebra of the Lie group G*.

Proposition. *If G is a matrix group, then the Lie algebra structure of* **g** *is given by*

$$[X,Y] = XY - YX$$

(where XY denotes the product of the matrices X, Y).

Sketch of the proof. We have $X^*f(g) = \frac{d}{dt}f(g \exp tX)|_0$. Hence,

$$(Y^*(X^*f))(e) = \tfrac{d}{ds}\tfrac{d}{dt}f(\exp sY \exp tX)|_0|_0.$$

If f is linear, the proposition follows from this. The general case can be deduced from the case where f is linear. ∎

Exercise:

(2.1) If $G = SO_3$ (and if **g** is identified with $\mathbf{R}^3$), show that we obtain Example (3) above.

The main significance of the Lie algebra (of a Lie group) is demonstrated by the next theorem:

Theorem. *There is a one to one correspondence between*

(1) connected Lie subgroups G of $GL_n\mathbf{R}$ (or $GL_n\mathbf{C}$), and

(2) Lie subalgebras V of $M_n\mathbf{R}$ (or $M_n\mathbf{C}$). ∎

The correspondence assigns to a Lie group G its Lie algebra **g**. (It follows from the definition of a Lie subgroup in Chapter 1 that this procedure is valid.) A proof of the theorem can be found in Varadarajan [1984], Theorem 2.5.2. There is a more general result, the "Fundamental Theorem of Lie Theory", which establishes a one to one correspondence between *arbitrary*

Lie groups (up to local isomorphism) and Lie algebras. Details of this may also be found in Varadarajan [1984], section 2.8.

Examples:

(1) $G = SO_n$, $V = \text{skew}_n \mathbf{R}$.

(2) For fixed $a, b \in \mathbf{R}$, with $(a, b) \neq (0, 0)$, let

$$G = \left\{ \begin{pmatrix} \exp \sqrt{-1}\, at & 0 \\ 0 & \exp \sqrt{-1}\, bt \end{pmatrix} \middle| t \in \mathbf{R} \right\}$$

$$V = \left\{ \begin{pmatrix} \sqrt{-1}\, at & 0 \\ 0 & \sqrt{-1}\, bt \end{pmatrix} \middle| t \in \mathbf{R} \right\}.$$

Observe that G is isomorphic to $S^1 = U_1$ if $b = 0$ or a/b is rational, and G is isomorphic to $\mathbf{R} = M_1 \mathbf{R}$ otherwise. (For comments on the topology of G, see the end of section II of Chapter I.)

(3) Let G be a Lie subgroup of $GL_n \mathbf{R}$. Hence, $\mathbf{g}$ is a (real) Lie subalgebra of $M_n \mathbf{R}$, and $\mathbf{g} \otimes \mathbf{C}$ is a (complex) Lie subalgebra of $M_n \mathbf{C}$. By the theorem, there exists a (complex) Lie subgroup G^c of $GL_n \mathbf{C}$ whose Lie algebra is $\mathbf{g} \otimes \mathbf{C}$. The complex group G^c is called the *complexification* of G. For example, $GL_n \mathbf{C}$ is the complexification of $GL_n \mathbf{R}$. *Warning: It is possible to have $G_1^c = G_2^c$ and $G_1 \neq G_2$!* For example, the complexification of U_n is also $GL_n \mathbf{C}$.

The correspondence between Lie groups and Lie algebras is extremely important because (very roughly speaking) it reduces the study of Lie groups to linear algebra. The Lie algebra is also a useful tool in the study of the differential geometry of Lie groups. As an example, we mention the following simple result.

A *Riemannian metric* $\langle \, , \, \rangle$ on a manifold X is by definition an inner product $\langle \, , \, \rangle_x$ on each tangent space $T_x X$ (and it is assumed that $\langle \, , \, \rangle_x$ depends smoothly on x). If $X = G$, a Lie group, then a *left-invariant* Riemannian metric is a Riemannian metric $\langle \, , \, \rangle$ such that $\langle X, Y \rangle_k = \langle DL_h X, DL_h Y \rangle_{hk}$ for all $X, Y \in T_k G$, and all $h, k \in G$.

Proposition. *There is a one to one correspondence between*

(1) left-invariant Riemannian metrics $\langle \, , \, \rangle$ on G, and

(2) inner products $\langle\langle \, , \, \rangle\rangle$ on $\mathbf{g}$. ∎

(Given $\langle \, , \, \rangle$, we define $\langle\langle \, , \, \rangle\rangle = \langle \, , \, \rangle_e$. Conversely, given $\langle\langle \, , \, \rangle\rangle$, we define $\langle X, Y \rangle_g = \langle\langle DL_{g^{-1}} X, DL_{g^{-1}} Y \rangle\rangle$, where $X, Y \in T_g G$.)

II. The adjoint representation.

Definition. *Let G be a Lie group. Let V be a vector space. A representation of G on V is a homomorphism $G \to GL(V)$.*

(The notation $GL(V)$ means the group of invertible linear transformations $T : V \to V$. For example, $GL(\mathbf{R}^n) = GL_n\mathbf{R}$.)

Definition. *Let G be a Lie group, and let $\mathbf{g}$ be the Lie algebra of G. The adjoint representation of G on $\mathbf{g}$ is the homomorphism*

$$\mathrm{Ad} : G \to GL(\mathbf{g}), \quad g \mapsto D(L_g \circ R_{g^{-1}})_e.$$

Alternatively:

$$\mathrm{Ad}(g)X = \tfrac{d}{dt} g \exp tX \, g^{-1}|_{t=0}.$$

Proposition. *If G is a matrix group, then $\mathrm{Ad}(A)X = AXA^{-1}$.*

Proof. $\mathrm{Ad}(A)X = \tfrac{d}{dt} A(\exp tX)A^{-1}|_0 = \tfrac{d}{dt} \exp tAXA^{-1}|_0 = AXA^{-1}$. ∎

There is a version of the adjoint representation for Lie algebras. First, a representation of a Lie algebra $\mathbf{g}$ on a vector space V is defined to be a Lie algebra homomorphism $\mathbf{g} \to \mathrm{End}(V)$. Next, the adjoint representation of $\mathbf{g}$ on $\mathbf{g}$ is defined to be the homomorphism

$$\mathrm{ad} = D(\mathrm{Ad})_e : \mathbf{g} \to \mathrm{End}(\mathbf{g}).$$

Proposition. *If G is a matrix group, then $\mathrm{ad}(X)Y = XY - YX$.*

Proof. $D(\mathrm{Ad})_e(X)Y = \tfrac{d}{dt} \mathrm{Ad}(\exp tX)Y|_0 = \tfrac{d}{dt}(\exp tX)Y(\exp -tX)|_0 = XY - YX$. ∎

(More generally, for arbitrary Lie groups, it can be shown that $\mathrm{ad}(X)Y = [X, Y]$.)

From now on, we shall only consider matrix groups! This is not a serious restriction, and it simplifies our calculations.

The adjoint representation is useful in studying the geometry of G. For example:

Proposition. *There is a one to one correspondence between*

(1) bi-invariant Riemannian metrics $\langle \, , \, \rangle$ on G (bi-invariant means left invariant and right invariant), and

(2) Ad-invariant inner products $\langle\langle\ ,\ \rangle\rangle$ on $\mathbf{g}$ (Ad-invariant means that $\langle\langle \mathrm{Ad}(g)X, \mathrm{Ad}(g)Y\rangle\rangle = \langle\langle X, Y\rangle\rangle$ for all $X, Y \in \mathbf{g}$ and all $g \in G$). ∎

This is proved in the same way as the proposition at the end of the last section.

By "averaging over G" any inner product on $\mathbf{g}$, and using the above correspondence, it is possible to prove the following existence result:

Proposition. *If G is a compact Lie group, then there exists a bi-invariant Riemannian metric on G.* ∎

For example, if $G = O_n$, an Ad-invariant inner product on $\mathrm{skew}_n \mathbf{R}$ is given by $\langle\langle A, B\rangle\rangle = -\operatorname{trace} AB$. By the proposition, we obtain a bi-invariant Riemannian metric on O_n.

A deeper property of the adjoint representation is that it measures the non-commutativity of G. For example, if G is abelian, then $\mathrm{Ad}(g)X = X$ for all $g \in G, X \in \mathbf{g}$. The next definition helps to clarify this idea.

Definition. *Let G be a compact Lie group. A Cartan subalgebra of $\mathbf{g}$ is a maximal abelian Lie subalgebra of $\mathbf{g}$.*

For example, let

$$
\mathbf{c}_n = \left\{ \begin{pmatrix} \sqrt{-1}\,x_1 & & & \\ & \sqrt{-1}\,x_2 & & \\ & & \ddots & \\ & & & \sqrt{-1}\,x_n \end{pmatrix} \ \middle|\ x_1, \ldots, x_n \in \mathbf{R} \right\}.
$$

Then $\mathbf{c}_n$ is a Cartan subalgebra of $\mathrm{skewHerm}_n \mathbf{C}$.

Theorem (Cartan). *Let G be a compact Lie group. Let $\mathbf{g}$ be the Lie algebra of G. Let $\mathbf{c}_1, \mathbf{c}_2$ be two Cartan subalgebras of $\mathbf{g}$. Then there exists some $g \in G$ such that $\mathrm{Ad}(g)\mathbf{c}_1 = \mathbf{c}_2$.* ∎

Corollary. *Let G be a compact Lie group. Let $\mathbf{g}$ be the Lie algebra of G. Let $\mathbf{c}$ be a Cartan subalgebra of $\mathbf{g}$. If $X \in \mathbf{g}$, then there exists some $g \in G$ such that $\mathrm{Ad}(g)X \in \mathbf{c}$.* ∎

This is a "diagonalization theorem". For example, let us take $G = U_n, \mathbf{g} = \mathrm{skewHerm}_n \mathbf{C}$, and $\mathbf{c} = \mathbf{c}_n$. Then we obtain the following familiar fact from linear algebra: If $X \in \mathrm{skewHerm}_n \mathbf{C}$, then there exists some $A \in U_n$ such that $AXA^{-1} \in \mathbf{c}_n$.

There is a version of Cartan's theorem for (compact, connected) Lie groups, where the concept of "Cartan subalgebra" is replaced by the concept of "maximal torus". As a corollary, we obtain diagonalization theorems for Lie groups. The appropriate definition is:

Definition. *Let G be a compact connected Lie group. A maximal torus of G is a maximal connected abelian subgroup of G.*

The name "torus" comes from the following well known fact:

Proposition. *If T is a connected abelian compact Lie group, then T is isomorphic to $S^1 \times \cdots \times S^1$ (r factors), for some integer r.* ∎

Example:

Let $G = U_n$. Let

$$
T_n = \left\{ \begin{pmatrix} z_1 & & & \\ & z_2 & & \\ & & \ddots & \\ & & & z_n \end{pmatrix} \;\middle|\; z_i \in \mathbf{C}, |z_i| = 1, i = 1, \dots, n \right\}.
$$

Then T_n is a maximal torus of U_n.

Cartan's theorem is:

Theorem (Cartan). *Let G be a compact connected Lie group. Let T_1, T_2 be two maximal tori of G. Then there exists some $g \in G$ such that $gT_1g^{-1} = T_2$.* ∎

Exercises:

(2.2) Let $\mathrm{Ad} : Sp_1 \to GL_3\mathbf{R}$ be the adjoint representation of Sp_1. Show that $\mathrm{Ad}(Sp_1) \cong SO_3$. Show that the kernel of Ad is $\{\pm I\}$. (It follows that SO_3 is diffeomorphic to three-dimensional real projective space $\mathbf{R}P^3$.)

(2.3) Find a Cartan subalgebra of $\mathrm{skew}_n \mathbf{R}$.

(2.4) Show that SO_4 is isomorphic to $Sp_1 \times Sp_1/\{\pm(1,1)\}$.

Bibliographical comments.

See the comments at the end of Chapter 3.

Chapter 3: Factorizations and homogeneous spaces

I. Decomposition of Lie groups.

If $\{v_1, \ldots, v_n\}$ is a basis of $\mathbf{R}^n$, the Gram-Schmidt procedure gives a new *orthonormal* basis $\{u_1, \ldots, u_n\}$. We may consider the vectors $v_1, \ldots, v_n$ to be the column vectors of a matrix $A \in GL_n\mathbf{R}$, and the vectors $u_1, \ldots, u_n$ to be the column vectors of a matrix $B \in O_n$. In terms of A and B, the Gram-Schmidt process gives a factorization $A = BC$, i.e.,

$$
\begin{pmatrix} | & & | \\ v_1 & \cdots & v_n \\ | & & | \end{pmatrix} = \begin{pmatrix} | & & | \\ u_1 & \cdots & u_n \\ | & & | \end{pmatrix} \begin{pmatrix} * & * & * & * & * \\ & * & * & * & * \\ & & * & * & * \\ & & & * & * \\ & & & & * \end{pmatrix}
$$

where C is an upper triangular matrix.

For example, if $n = 2$, then u_1, u_2 are given by

$$u_1 = v_1/|v_1| = \alpha v_1, \text{ say, and}$$
$$u_2 = (v_2 - \langle v_2, u_1 \rangle u_1)/|v_2 - \langle v_2, u_1 \rangle u_1| = \beta v_2 + \gamma v_1, \text{ say.}$$

This may be written

$$B = A \begin{pmatrix} \alpha & \gamma \\ 0 & \beta \end{pmatrix},$$

or

$$A = B \begin{pmatrix} \alpha^{-1} & -\gamma/(\alpha\beta) \\ 0 & \beta^{-1} \end{pmatrix}.$$

It follows that we have a decomposition of Lie groups, namely

$$GL_n\mathbf{R} = O_n \Delta_n,$$

where Δ_n is the subgroup of upper triangular matrices in $GL_n\mathbf{R}$. We have

$$O_n \cap \Delta_n = Z_n$$

where Z_n is the (finite) group of matrices of the form

$$\begin{pmatrix} \pm 1 & & & & \\ & \pm 1 & & & \\ & & \pm 1 & & \\ & & & \pm 1 & \\ & & & & \pm 1 \end{pmatrix}.$$

Similarly, using an oriented basis, we obtain the decomposition

$$SL_n\mathbf{R} = SO_n\Delta_n, \quad SO_n \cap \Delta_n = Z_n.$$

Here we are using the (rather dangerous, but convenient) convention that Δ_n denotes the group of upper triangular matrices in $SL_n\mathbf{R}$ (not $GL_n\mathbf{R}$!), and similarly for Z_n.

If we use complex vectors instead of real vectors, we obtain the decomposition

$$GL_n\mathbf{C} = U_n\Delta_n, \quad U_n \cap \Delta_n = T_n$$

where Δ_n is the group of upper triangular matrices in $GL_n\mathbf{C}$, and T_n is the group of diagonal matrices in U_n.

We also have a decomposition

$$SL_n\mathbf{C} = SU_n\Delta_n, \quad SU_n \cap \Delta_n = T_n$$

where Δ_n and T_n denote the upper triangular and diagonal subgroups of $SL_n\mathbf{C}$ and SU_n, respectively.

There is a general decomposition theorem for Lie groups, called the *Iwasawa decomposition* (see Helgason [1978]):

Theorem (Iwasawa decomposition). *Let G be a compact connected Lie group. Then there is a decomposition*

$$G^c = GAN, \quad G \cap A = A \cap N = G \cap N = \{e\}$$

where A is abelian and N is nilpotent. ∎

Let us examine how this applies to the examples $G = U_n, SU_n$ above. (For the examples O_n, SO_n, a slightly different theorem is available – see section I of Chapter 7.) In the case $G = U_n$, we have seen that $GL_n\mathbf{C} = U_n\Delta_n$. It is easy to see that $\Delta_n = T_nA_nN_n$, where A_n is the group of real diagonal matrices with positive entries, and where N_n is the group of matrices in Δ_n of the form

$$\begin{pmatrix} 1 & * & * & * & * \\ & 1 & * & * & * \\ & & 1 & * & * \\ & & & 1 & * \\ & & & & 1 \end{pmatrix}.$$

Since $T_n \subseteq U_n$, we have $GL_n\mathbf{C} = U_nA_nN_n$, and this gives the Iwasawa decomposition of $GL_n\mathbf{C}$.

(A Lie group is said to be nilpotent if its Lie algebra is nilpotent; a (matrix) Lie algebra is said to be nilpotent if each element X is a nilpotent matrix, i.e., $X^k = 0$ for some k. The Lie algebra of the group N_n consists of all strictly upper triangular matrices.)

Exercise:

(3.1) Find the Iwasawa decomposition explicitly, in the case of the group $GL_2\mathbf{C}$.

Next we describe a geometrical version of the decomposition $GL_n\mathbf{C} = U_n\Delta_n$.

Definition.

(1) $\mathrm{F} = \mathrm{F}_{1,2,\dots,n-1}(\mathbf{C}^n) =$

$\{\{0\} \subseteq E_1 \subseteq E_2 \subseteq \dots \subseteq E_{n-1} \subseteq \mathbf{C}^n \mid E_i$ *a subspace of* $\mathbf{C}^n$ *of dimension* $i\}$.

(2) $\mathrm{F}' = \{(L_1,\dots,L_n) \mid L_1,\dots,L_n$ *are orthogonal lines in* $\mathbf{C}^n\}$.

Proposition.

(1) $\mathrm{F} \cong GL_n\mathbf{C}/\Delta_n$.

(2) $\mathrm{F}' \cong U_n/T_n$.

Proof. (1) The group $GL_n\mathbf{C}$ acts naturally on F. The isotropy subgroup at the point

$$\{0\} \subseteq \mathbf{C} \subseteq \mathbf{C}^2 \subseteq \dots \subseteq \mathbf{C}^{n-1} \subseteq \mathbf{C}^n \quad (\in \mathrm{F})$$

is Δ_n. We claim that the action is transitive. To see this, let $v_1,\dots,v_n$ be a basis of $\mathbf{C}^n$ such that $E_i = \mathrm{Span}\{v_1,\dots,v_i\}$. Then the matrix

$$\begin{pmatrix} | & & | \\ v_1 & \cdots & v_n \\ | & & | \end{pmatrix}$$

defines an element of $GL_n\mathbf{C}$ which maps $\{0\} \subseteq \mathbf{C} \subseteq \mathbf{C}^2 \subseteq \dots \subseteq \mathbf{C}^{n-1} \subseteq \mathbf{C}^n$ to $\{0\} \subseteq E_1 \subseteq E_2 \subseteq \dots \subseteq E_{n-1} \subseteq \mathbf{C}^n$. (2) The group U_n acts naturally on F'. Let $e_1,\dots,e_n$ be an orthonormal basis of $\mathbf{C}^n$. The isotropy subgroup at the point

$$(\mathbf{C}e_1,\dots,\mathbf{C}e_n) \quad (\in \mathrm{F}')$$

is T_n. The action is transitive, by an argument similar to the argument for (1). ∎

Corollary. $GL_n\mathbf{C} = U_n\Delta_n$.

Proof. The map $F' \to F$, given by

$$(L_1, \ldots, L_n) \mapsto \{0\} \subseteq E_1 \subseteq E_2 \subseteq \ldots \subseteq E_{n-1} \subseteq \mathbf{C}^n, \quad E_i = L_1 \oplus \cdots \oplus L_i,$$

is bijective. (In fact, it is a diffeomorphism.) This can be identified with the natural map

$$U_n/T_n \to GL_n\mathbf{C}/\Delta_n.$$

From the surjectivity of this map, we obtain $GL_n\mathbf{C} = U_n\Delta_n$. ∎

The manifold F (or F') is called a *flag manifold*.

More generally, the following theorem holds:

Theorem. *Let G be a compact connected Lie group. Let T be a maximal torus of G. Then there is a subgroup B of G^c such that $G/T \cong G^c/B$.* ∎

The group B is called a *Borel subgroup* of G^c. The manifold G^c/B (or G/T) is called a *generalized flag manifold*. This theorem is a consequence of the proof of the general Iwasawa decomposition (esssentially, we have $B = TAN$).

II. Homogeneous spaces.

The following result is elementary.

Proposition. *Let G be a Lie group. Let H be a closed Lie subgroup of G. Then the coset space $G/H = \{gH \mid g \in G\}$ has the structure of a (smooth) manifold.* ∎

We call G/H a *homogeneous space*. If $g \in G$, we sometimes write $[g] = gH$, and $o = [e] = eH$. We have $\mathbf{h} \subseteq \mathbf{g}$.

Examples:

(1) The flag manifolds F (or F'). The generalized flag manifolds G^c/B (or G/T).

(2) The spheres S^n.

(3) The Grassmannians $\mathrm{Gr}_m(\mathbf{R}^n), \mathrm{Gr}_m(\mathbf{C}^n), \mathrm{Gr}_m(\mathbf{H}^n)$.

Exercise:

(3.2) Find suitable groups G, H, for Examples (2) and (3) above.

Homogeneous spaces are more general than Lie groups, but they have some similar properties. For example, there is a close relationship between a homogeneous space G/H and its tangent space T_oG/H at o. We have an isomorphism

$$\mathbf{g}/\mathbf{h} \cong T_oG/H,$$

given by

$$[X] \mapsto \tfrac{d}{dt}[\exp tX]|_0.$$

For any $X \in \mathbf{g}$, we can define a vector field X^* on G/H by:

$$X^*_{gH} = \tfrac{d}{dt}(\exp tX)gH|_0 = \tfrac{d}{dt}[(\exp tX)g]|_0.$$

(Beware: This differs from the formula for Lie groups in Chapter 2!)

The adjoint representation $\mathrm{Ad} = \mathrm{Ad}^G$ of a Lie group G is related to the *isotropy representation* of a homogeneous space:

Definition. *Let G/H be a homogeneous space. The isotropy representation of H on T_oG/H is the homomorphism*

$$\mathrm{Ad}^{G/H} : H \to GL(T_oG/H), \quad \mathrm{Ad}^{G/H}(h)X = DL_h(X)$$

where $L_h : G/H \to G/H$ is the map $L_h([g]) = [hg]$.

Exercise:

(3.3) Check that $\mathrm{Ad}^{G/H}$ is well defined.

Next, we explain the precise relationship between Ad^G and $\mathrm{Ad}^{G/H}$.

Definition. *The homogeneous space G/H is reductive if there exists a subspace $\mathbf{m} \subseteq \mathbf{g}$ such that $\mathbf{g} = \mathbf{h} \oplus \mathbf{m}$ and $\mathrm{Ad}(h)\mathbf{m} \subseteq \mathbf{m}$ for all $h \in H$.*

(If G is compact, then G/H is reductive, because we can take $\mathbf{m} = \mathbf{h}^\perp$, with respect to an Ad-invariant inner product on $\mathbf{g}$.)

Proposition. *Assume that G/H is reductive. Let $h \in H$, and let $X \in \mathbf{h}$, $Y \in \mathbf{m}$. Then we have*

$$\mathrm{Ad}^G(h)(X,Y) = (\mathrm{Ad}^H(h)X, \mathrm{Ad}^{G/H}(h)Y).$$

Proof. It suffices to prove this for (i) $X = 0$, and (ii) $Y = 0$. The result is obvious for case (ii). For case (i), we use the fact that Y (in $\mathbf{m}$) corresponds to $\tfrac{d}{dt}[\exp tY]|_0$ (in $T_o(G/H)$). Hence, we must show that $\tfrac{d}{dt}[\exp t\,\mathrm{Ad}^G(h)Y]|_0$ is equal to $\mathrm{Ad}^{G/H}(h)\tfrac{d}{dt}[\exp tY]|_0$. We have:

$$\mathrm{Ad}^{G/H}(h)\tfrac{d}{dt}[\exp tY]|_0 = \tfrac{d}{dt}[h\exp tY]|_0$$
$$= \tfrac{d}{dt}[h(\exp tY)h^{-1}]|_0$$
$$= \tfrac{d}{dt}[\exp t\,\mathrm{Ad}^G(h)Y]|_0$$

as required. ■

In the language of representation theory, we say that the restriction of Ad^G to H splits into the sum $\mathrm{Ad}^H \oplus \mathrm{Ad}^{G/H}$.

Exercise: (This exercise assumes some knowledge of representation theory.)

(3.4) Identify $\mathrm{Ad}^G, \mathrm{Ad}^H, \mathrm{Ad}^{G/H}$ in terms of "standard representations", in Examples (1),(2),(3) above.

Finally, we mention the following description of certain Riemannian metrics on a reductive homogeneous space G/H:

Proposition. *There is a one to one correspondence between*

(1) invariant Riemannian metrics $\langle\ ,\ \rangle$ on G/H (invariant means that $\langle X, Y \rangle = \langle DL_h X, DL_h Y \rangle$ for all $X, Y \in T_{[k]}G$, and all $h, k \in G$), and

(2) $\mathrm{Ad}^{G/H}$-invariant inner products $\langle\langle\ ,\ \rangle\rangle$ on $\mathbf{m}$ ($\mathrm{Ad}^{G/H}$-invariant means that $\langle\langle \mathrm{Ad}^{G/H}(h)X, \mathrm{Ad}^{G/H}(h)Y \rangle\rangle = \langle\langle X, Y \rangle\rangle$ for all $X, Y \in \mathbf{m}$ and all $h \in H$). ■

This is similar to the propositions in Chapter 2, concerning Riemannian metrics on G.

Bibliographical comments for Chapters 1-3.

The literature on Lie groups and Lie algebras is very well developed, and the reader will experience no difficulty in finding books on the subject. We just mention some particularly useful sources for the topics of Chapters 1-3. Accessible introductions are given in Warner [1984], the first chapter of Bröcker and tom Dieck [1985], and the first two chapters of Adams [1969]. An excellent reference for all the material of Chapters 1-3 is the book by Carter *et al.* [1995]. (There is a concise discussion of the Fundamental Theorem of Lie Theory on pages 75-81.) For a more detailed study see Helgason [1978]; Postnikov [1986]; Varadarajan [1984].

Chapter 4: Hamilton's equations and Hamiltonian systems

I. Hamilton's equations.

Newton's equations for the motion of a particle of mass m, in a field with potential $V = V(q)$, are

$$(*) \qquad m\ddot{q}_j = -\frac{\partial V}{\partial q_j},$$

where $q = (q_1, \ldots, q_n) \in \mathbf{R}^n$ denotes the position of the particle. Let us introduce

$$p_j = m\dot{q}_j \quad \text{and} \quad H(p, q) = \frac{1}{2m} \sum_{j=1}^{n} p_j^2 + V(q),$$

the "momentum" and "total energy", respectively. Then we obtain Hamilton's equations for (p, q):

$$(**) \qquad \dot{q}_j = \frac{\partial H}{\partial p_j}, \quad \dot{p}_j = -\frac{\partial H}{\partial q_j}.$$

Equations $(*),(**)$ are equivalent, but $(**)$ has various advantages. For example, $(**)$ contains only first order derivatives, whereas $(*)$ contains second order derivatives. Equation $(**)$ is "more symmetrical" than $(*)$, perhaps.

If we introduce

$$x = \begin{pmatrix} p^t \\ q^t \end{pmatrix} \quad \text{and} \quad J = \begin{pmatrix} 0 & -I \\ I & 0 \end{pmatrix},$$

then $(**)$ becomes:

$$(* * *) \qquad \dot{x} = J(\nabla H)_x.$$

In this equation, the gradient ∇H of $H : \mathbf{R}^n \times \mathbf{R}^n \to \mathbf{R}$ is defined (as usual) in the following way: $\langle\langle (\nabla H)_V, W \rangle\rangle = DH_V(W)$ for all $V, W \in \mathbf{R}^n \times \mathbf{R}^n$, where $\langle\langle \ , \ \rangle\rangle$ is the standard inner product on $\mathbf{R}^n \times \mathbf{R}^n$.

Equation $(* * *)$ has the following generalization:

(1) Let M be a (smooth) manifold.

(2a) Let $\langle \ , \ \rangle$ be a Riemannian metric on M (i.e., a smoothly varying inner product $\langle \ , \ \rangle_m$ on each tangent space $T_m M$).

(2b) Let J be an almost complex structure on M (i.e., a smoothly varying automorphism J_m of each tangent space $T_m M$, such that $(J_m)^2 = -I$).

(3) Let $H : M \to \mathbf{R}$ be any (smooth) function.

Then we may consider equation (∗∗∗), in this situation. It is a first order differential equation for a path $x : \mathbf{R} \to M$.

More generally still, we may replace (2a) and (2b) by:

(2) Let ω be a non-degenerate 2-form on M (non-degenerate means that if $\omega(X,Y) = 0$ for all Y, then $X = 0$).

Given (1), (2) and (3), we may consider the equation

$$\dot{x} = V_x$$

where V is the vector field such that $\omega(V, \ ) = dH$. This equation is called a *Hamiltonian system*; H is called the *Hamiltonian function* (of the system), and V is called the *Hamiltonian vector field* (of the system). Of course, this equation is simply the equation for an integral curve of the vector field V. If we define ω by $\omega(X,Y) = \langle X, JY \rangle$ in the previous situation (with (2a) and (2b)), then we have $V = J\nabla H$.

It is usual in the definition of a Hamiltonian system to insist that ω is closed, i.e., that $d\omega = 0$. (This is the case in the above example.) A non-degenerate closed 2-form is called a *symplectic form*, or a *symplectic structure* (on M).

II. Example: A height function on an Ad-orbit.

We discuss an important example of a Hamiltonian system. First we must define M, ω, and H.

(1) *Definition of M.*

Let G be a compact Lie group. Let $P \in \mathbf{g}$.

Definition. $M_P = \mathrm{Ad}(G)P = \{\mathrm{Ad}(g)P \mid g \in G\} \subseteq \mathbf{g}$.

This M_P is called an "adjoint orbit" (or "Ad-orbit"). If $H_P = \{g \in G \mid \mathrm{Ad}(g)P = P\}$, i.e., the isotropy subgroup at P, then $M_P \cong G/H_P$, so M_P is an example of a homogeneous space. The point $P \in M_P$ corresponds to the point $o = [e] \in G/H_P$.

Examples:

(1) $G = U_n$, $P = \begin{pmatrix} \sqrt{-1}\,x_1 & & & \\ & \sqrt{-1}\,x_2 & & \\ & & \cdots & \\ & & & \sqrt{-1}\,x_n \end{pmatrix}$,

with $x_1, \ldots, x_n$ distinct real numbers. In this case, $H_P = T_n$ and so $M_P = F_{1,2,\ldots,n-1}(\mathbf{C}^n)$.

$$(2)\ G = U_n,\ P = \begin{pmatrix} \sqrt{-1}\,x_1 & & & \\ & \sqrt{-1}\,x_2 & & \\ & & \cdots & \\ & & & \sqrt{-1}\,x_n \end{pmatrix},$$

with $x_1 = \cdots = x_k = x$, $x_{k+1} = \cdots = x_n = y$, where x, y are distinct real numbers. In this case, $H_P = U_k \times U_{n-k}$ and so $M_P = \mathrm{Gr}_k(\mathbf{C}^n)$.

Exercise:

(4.1) Find some more examples!

As in the case of a flag manifold (see Chapter 3), it can be shown that $G/H_P \cong G^c/G_P$, for a suitable subgroup G_P of G^c.

The next proposition gives us a description of the "abstract" tangent space of M_P:

Proposition. *The isotropy subgroup H_P is equal to*

$$C(S_P) = \{g \in G \mid ghg^{-1} = h \text{ for all } h \in S_P\}$$

where $S_P = \overline{\exp \mathbf{R}P}$. The Lie algebra of H_P is given by $\mathbf{h}_P = \{X \in \mathbf{g} \mid [P, X] = 0\}$. ■

We have the decomposition $\mathbf{g} = \mathbf{h}_P \oplus \mathbf{m}_P$, where $\mathbf{m}_P = (\mathbf{h}_P)^\perp$. (The orthogonal complement is taken with respect to an Ad-invariant inner product $(\ ,\ )$ on $\mathbf{g}$.) Thus, the tangent space (at P) of M_P is $T_P(M_P) \cong \mathbf{m}_P = \{X \in \mathbf{g} \mid [P, X] = 0\}^\perp$.

On the other hand, we can use the embedding $M_P \subseteq \mathbf{g}$ to give another description of the tangent space to M_P:

$$\begin{aligned} T_P(M_P) &= \{\tfrac{d}{dt}\,\mathrm{Ad}(\exp tX)P|_0 \mid X \in \mathbf{g}\} \\ &= \{\tfrac{d}{dt}(\exp tX)P(\exp -tX)|_0 \mid X \in \mathbf{g}\} \\ &= \{XP - PX \mid X \in \mathbf{g}\} \\ &= \{[X, P] \mid X \in \mathbf{g}\}. \end{aligned}$$

Let $A_P : \mathbf{g} \to \mathbf{g}$ be defined by $A_P(X) = [P, X]$. Then we have two descriptions of $T_P(M_P)$, namely (i) $(\mathrm{Ker}\, A_P)^\perp$, and (ii) $\mathrm{Im}\, A_P$. It turns out that these are the same:

Proposition. $\mathrm{Im}\, A_P = (\mathrm{Ker}\, A_P)^\perp$.

The proof is given in the following exercises.

Exercises:

(4.2) Show that $([X,Y],Z) = (X,[Y,Z])$ for all $X,Y,Z \in \mathbf{g}$. (Hint: Use

(a) $[X,Y] = \frac{d}{dt}\operatorname{Ad}(\exp tX)Y|_0$, and

(b) $(\operatorname{Ad}(g)X,\operatorname{Ad}(g)Y) = (X,Y)$.)

(4.3) Show that $\operatorname{Ker} A_P \perp \operatorname{Im} A_P$. (Hint: Use (4.2).) ∎

The embedding $M_P = \{\operatorname{Ad}(g)P \mid g \in G\} \subseteq \mathbf{g}$ is often useful for computational purposes. For example, we can obtain a simple formula for the vector field X^* on M_P:

$$
\begin{aligned}
X^*_{\operatorname{Ad}(g)P} &= \tfrac{d}{dt}\operatorname{Ad}(\exp tX)\operatorname{Ad}(g)P|_0 \\
&= \tfrac{d}{dt}(\exp tX)\operatorname{Ad}(g)P(\exp -tX)|_0 \\
&= [X,\operatorname{Ad}(g)P].
\end{aligned}
$$

(2) *Definition of ω.*

For any reductive homogeneous space G/H, there is a one to one correspondence between

(1) invariant 2-forms ω on G/H

(2) $\operatorname{Ad}^{G/H}$-invariant skew-symmetric bilinear forms Ω on $\mathbf{m}$.

This is similar to the corresponding statement for Riemannian metrics, in Chapter 3. If G is a compact Lie group, and $H = H_P$, we shall define Ω as follows:

Definition. *For any $X,Y \in \mathbf{m}$, $\Omega(X,Y) = (P,[X,Y])$.*

(Here, $(\ ,\ )$ is an Ad-invariant inner product on $\mathbf{g}$, as usual.)

Exercises:

(4.4) Show that $\Omega(\operatorname{Ad}(h)X,\operatorname{Ad}(h)Y) = \Omega(X,Y)$ for all $h \in H_P$.

(4.5) Let ω be the 2-form corresponding to Ω. Show that $\omega(X^*_{gH},Y^*_{gH}) = \Omega(X,Y)$ for all $X,Y \in \mathbf{g}$.

(4.6) Show that ω is non-degenerate.

(4.7) Show that $d\omega = 0$.

These exercises show that ω is indeed a symplectic form on M_P.

(3) *Definition of H.*

Definition. *Let $Q \in \mathbf{g}$. Define $H^Q : M_P \to \mathbf{R}$ by $H^Q(X) = (X,Q)$.*

(Of course, we also have a function $\hat{H}^Q : \mathbf{g} \to \mathbf{R}$ defined by $\hat{H}^Q(X) = (X,Q)$, and $H^Q = \hat{H}^Q|_{M_P}$.) We call H^Q a *height function* on M_P.

(Another natural function is the *distance function*
$$K^Q(\mathrm{Ad}(g)P) = |\,\mathrm{Ad}(g)P - Q|^2.$$
But this is essentially the same as H^Q, because $K^Q = a + bH^Q$ for some constants a, b.)

This completes the definition of M, ω, and H.

We now have a Hamiltonian system $\dot{x} = V_x$. In this example it is possible to find a complex structure J and a Riemannian metric $\langle\,,\,\rangle$ on M_P, such that $V = J\nabla H^Q$. These may be defined explicitly, by using the *roots* of G. We quote the following result without proof:

Proposition. $\nabla H^Q = JQ^*.$ ∎

Our Hamiltonian system is $\dot{x} = J(\nabla H^Q)_x$. From the last proposition we have $J(\nabla H^Q)_x = -Q_x^* = [x, Q]$. Hence our differential equation for $x : \mathbf{R} \to M_P$ is
$$\dot{x} = [x, Q].$$
It is easy to find the (unique) solution to this differential equation:

Theorem. *The solution of the differential equation*
$$\dot{x} = [x, Q], \quad x(0) = P$$
is given by $x(t) = \mathrm{Ad}(\exp -tQ)P.$

Proof. It suffices to show that $x(t) = \mathrm{Ad}(\exp -tQ)P$ satisfies the differential equation. We have
$$
\begin{aligned}
\dot{x}(t) &= \tfrac{d}{ds}x(t + s)|_0 \\
&= \tfrac{d}{ds}\mathrm{Ad}(\exp -(t + s)Q)P|_0 \\
&= \mathrm{Ad}(\exp -tQ)\tfrac{d}{ds}\mathrm{Ad}(\exp -sQ)P|_0 \\
&= \mathrm{Ad}(\exp -tQ)[-Q, P] \\
&= \mathrm{Ad}(\exp -tQ)[P, Q] \\
&= [\mathrm{Ad}(\exp -tQ)P, \mathrm{Ad}(\exp -tQ)Q] \\
&= [x(t), Q].\ \blacksquare
\end{aligned}
$$

Before we leave this example, let us consider the differential equation for the vector field ∇H^Q:
$$\dot{x} = (\nabla H^Q)_x \quad \text{i.e.,} \quad \dot{x} = J[x, Q].$$
(This is not a Hamiltonian system.) The solutions to this equation are slightly more complicated, but it is possible to find them explicitly by using the Iwasawa decomposition $G^c = GAN$ (of Chapter 3). We shall not give the proof, because we have not defined J precisely. In the following statement, we shall write $g = g_u g_a g_n$ for the factorization of $g \in G^c$.

Theorem. *The solution of the differential equation*

$$\dot{x} = J[x, Q], \quad x(0) = P$$

is given by $x(t) = \mathrm{Ad}(\exp \sqrt{-1}\, tQ)_u P.$ ∎

Exercises:

(4.8) Let $G = SU_2$, and let

$$P = \begin{pmatrix} \sqrt{-1} & 0 \\ 0 & \sqrt{-1} \end{pmatrix}, \quad Q = \begin{pmatrix} 0 & 1 \\ -1 & 0 \end{pmatrix}.$$

Show that $M_P \cong S^2$. Determine explicitly the embedding $M_P \subseteq \mathbf{g} \cong \mathbf{R}^3$.

(4.9) Show that X (in M_P) is a critical point of $H^Q : M_P \to \mathbf{R}$ if and only if $[Q, X] = 0$.

Bibliographical comments.

The modern theory of classical mechanics (in terms of symplectic manifolds) is discussed in the book by Arnold [1978]. A broader treatment of symplectic manifolds in physics is given in the book by Guillemin and Sternberg [1984].

The example of section II appears infrequently, but in various guises, in the literature. The height function H^Q was studied from the point of view of Morse theory in Bott [1956]. It turns out that H^Q is a "perfect Morse function"; this allows one to obtain information on the homology of M_P. (The classical reference for Morse theory is the book by Milnor [1963]. A brief summary of the Morse theory of H^Q is given in the Appendix of Guest and Ohnita [1993].) A more general (symplectic) point of view was taken in Frankel [1959], and subsequently in Atiyah [1982].

From the point of view of Morse theory, it is natural to consider the "gradient flow" curve $t \mapsto \mathrm{Ad}(\exp \sqrt{-1}\, tQ)_u P$; its closure is a curve of finite length from a (higher) critical point of H^Q to a (lower) critical point. From the Hamiltonian point of view it is natural to consider the "orthogonal" curve $t \mapsto \mathrm{Ad}(\exp tQ)P$; the closure of this is the orbit of P under the torus S_Q. (We assume here that P is not a critical point of H^Q, i.e., not a fixed point of the action of S_Q, otherwise both curves are just the point P itself.) The reader is advised to contemplate these two curves in the situation of Exercise (4.8). There, the critical points of H^Q are just the antipodal points $\pm Q$. Thinking of $\pm Q$ as the north and south poles, the first curve is a "line of longitude" between the poles, and the second curve is a "circle of latitude".

Chapter 5: Lax equations

I. Flows on adjoint orbits.

An equation of the form

$$\dot{L} = [L, M]$$

is called a *Lax equation*. In the last chapter we considered an example, namely $\dot{x} = [x, Q]$, where $x : \mathbf{R} \to \mathbf{g}$, $Q \in \mathbf{g}$, and G is a compact Lie group. In this chapter we shall consider the following more general example:

$$(*) \qquad \dot{x} = [x, y], \quad x(0) = V$$

where $x, y : \mathbf{R} \to \mathbf{g}$, and G is any Lie group. The key to solving this equation (for certain y, at least) is the geometrical property established in the next proposition. We shall denote the adjoint orbit of V by O_V, i.e., $O_V = \{\mathrm{Ad}(g)V \mid g \in G\} = \mathrm{Ad}(G)V$.

Proposition. *If x is a solution of $(*)$, then we have $x(t) \in O_V$ for all t.*

Proof. We proved in Chapter 4 that $T_x O_V = \{[X, x] \mid X \in \mathbf{g}\}$, if G is compact. The same proof is valid if G is non-compact. Therefore, $\dot{x} (= [x, y]) \in T_x O_x$. It can be deduced from this that $x(t) \in O_V$ for all t. ■

Therefore, we may write $x(t) = \mathrm{Ad}\, u(t)V (= u(t)Vu(t)^{-1})$, for some $u : (-\epsilon, \epsilon) \to G$. (This is justified by the fact that the natural map $G \to G/H = O_V$ is a locally trivial fibre bundle.) Differentiating the equation $x = uVu^{-1}$, we obtain

$$
\begin{aligned}
\dot{x} &= \dot{u}Vu^{-1} - uVu^{-1}\dot{u}u^{-1} \\
&= (\dot{u}u^{-1})(uVu^{-1}) - (uVu^{-1})(\dot{u}u^{-1}) \\
&= [\dot{u}u^{-1}, uVu^{-1}] \\
&= [x, -\dot{u}u^{-1}].
\end{aligned}
$$

Comparing this with $(*)$, we see that $(*)$ is equivalent to the following equation:

$$(**) \qquad \dot{u}u^{-1} = -y, \quad u(0) = e.$$

Thus, the "change of variable" suggested by the above geometrical property leads to a simplification of the equation.

If y is constant (i.e., independent of t), then the equation $(**)$ has the obvious solution $u(t) = \exp -ty$. We therefore obtain the solution

$x(t) = \mathrm{Ad}\exp(-ty)V$ of equation (∗). This explains the formula given in the last chapter!

A slightly more general example may be obtained by "working backwards". Suppose that we have a decomposition $G = G_1 G_2$, where G_1, G_2 are subgroups of G such that $G_1 \cap G_2 = \{e\}$. If $g \in G$, we may write $g = g_1 g_2$, where $g_1 \in G_1$, $g_2 \in G_2$. Let us consider $u = g_1$, where $g(t) = \exp tW$, for some $W \in \mathbf{g}$ (W is independent of t). We shall find an equation for which u is a solution. We have[1]:

$$
\begin{aligned}
W &= \dot{g}g^{-1} \\
&= (\dot{g}_1 g_2 + g_1 \dot{g}_2)g_2^{-1}g_1^{-1} \\
&= \dot{g}_1 g_1^{-1} + g_1(\dot{g}_2 g_2^{-1})g_1^{-1}.
\end{aligned}
$$

Now, we have $\mathbf{g} = \mathbf{g}_1 \oplus \mathbf{g}_2$, and $\dot{g}_1 g_1^{-1} \in \mathbf{g}_1$, $\dot{g}_2 g_2^{-1} \in \mathbf{g}_2$. *If we assume that* $\mathrm{Ad}\, G_1(\mathbf{g}_2) \subseteq \mathbf{g}_2$, then $\dot{u}u^{-1} = \dot{g}_1 g_1^{-1} = \pi_1 W$. (Here $\pi_1 W$ means W_1, where $W = (W_1, W_2)$.) Bearing in mind the relationship between (∗) and (∗∗), we have just proved the following fact:

Proposition. *Let* $G = G_1 G_2$, *where* G_1, G_2 *are subgroups of* G *such that* $G_1 \cap G_2 = \{e\}$. *Assume that* $\mathrm{Ad}\, G_1(\mathbf{g}_2) \subseteq \mathbf{g}_2$. *Let* $W \in \mathbf{g}$. *Then the solution of the differential equation*

$$
\dot{x} = [x, \pi_1 W], \quad x(0) = V
$$

is $x(t) = \mathrm{Ad}(\exp -tW)_1 V$. ∎

This result is closely related to the example of Chapter 4. The latter may be obtained by taking $G_1 = G, G_2 = \{e\}$. Alternatively we may regard the proposition as a special case of that example, with $P = V$, $Q = \pi_1 W$. The solution given by Chapter 4 is then $x(t) = \mathrm{Ad}(\exp -t\pi_1 W)V$ – which appears to be different from the formula of the proposition. However, the two formulae are in fact the same, as $(\exp -tW)_1 = \exp -t\pi_1 W$ here. This follows from the condition $\mathrm{Ad}\, G_1(\mathbf{g}_2) \subseteq \mathbf{g}_2$ and the Baker-Campbell-Hausdorff formula.

If we consider $u(t) = (\exp tW)_1^{-1}$, then we obtain another, less trivial, example:

---

[1] Two technical points should be mentioned. First, the notation $\dot{g}_1$ means $\frac{d}{dt}(g_1)$, and the notation g_1^{-1} means $(g_1)^{-1}$, i.e., factorization always occurs first. Second, the smoothness of $t \mapsto g_1(t), g_2(t)$ follows from the smoothness of $t \mapsto g(t)$, as the hypothesis on G_1, G_2 implies that the map $G_1 \times G_2 \to G$, $(a, b) \mapsto ab$, is a diffeomorphism.

Proposition. *Let $G = G_1 G_2$, where G_1, G_2 are subgroups of G such that $G_1 \cap G_2 = \{e\}$. Then the solution of the differential equation*

$$\dot{x} = [x, \pi_1 x], \quad x(0) = V$$

is $x(t) = \mathrm{Ad}(\exp tV)_1^{-1} V$.

Proof. The calculation is similar to that in the previous case, but we shall give it in full in order to emphasize how elementary it is. To simplify notation, we write $g(t) = \exp tW$, $u = g_1$ and $x = u^{-1} V u$. Differentiating the last equation, we obtain:

(†) $\dot{x} = [x, u^{-1} \dot{u}].$

As in the previous proposition, we have

$$W = \dot{g}_1 g_1^{-1} + g_1 (\dot{g}_2 g_2^{-1}) g_1^{-1},$$

hence

$$g_1^{-1} W g_1 = g_1^{-1} \dot{g}_1 + \dot{g}_2 g_2^{-1},$$

and so $g_1^{-1} \dot{g}_1 = \pi_1 (g_1^{-1} W g_1)$. If we take $W = V$ now, then we have

(††) $g_1^{-1} \dot{g}_1 = \pi_1 x.$

Comparing (†) and (††), we see that x satisfies the equation $\dot{x} = [x, \pi_1 x]$, as required. ■

II. Example: The Toda lattice.

We shall consider an important example of the above theory, namely the *Toda lattice*. It is a Hamiltonian system which describes the motion of n particles[2] moving in a straight line, with "exponential interactions". (Mathematically, this is equivalent to a problem in which a single particle moves in $\mathbf{R}^n$, so we may use the framework of Chapter 4.)

Let the positions of the particles at time t (in $\mathbf{R}$) be $q_1(t), \ldots, q_n(t)$, respectively. We assume that each particle has mass 1. The momentum of the i-th particle at time t is therefore $p_i = \dot{q}_i$. The Hamiltonian function is defined to be

$$H = \tfrac{1}{2} \sum_{j=1}^{n} p_j^2 + \sum_{j=1}^{n-1} e^{2(q_j - q_{j+1})}.$$

---

[2]This is the popular interpretation of the Toda lattice. However, in Toda's original interpretation, q_i was regarded as the angular parameter of a particle which moves on a circle. Adjacent particles are connected by springs. This accounts for the appearance of $q_i - q_{i+1}$ rather than $|q_i - q_{i+1}|$ in the potential. I am grateful to Professor B. Kostant for this information.

The Hamiltonian system

$$\dot{q}_j = \frac{\partial H}{\partial p_j}, \quad \dot{p}_j = -\frac{\partial H}{\partial q_j}$$

is

$$\dot{q}_j = p_j \quad j = 1, \ldots, n$$

$$\dot{p}_1 = -2e^{2(q_1 - q_2)}$$
$$\dot{p}_j = -2e^{2(q_j - q_{j+1})} + 2e^{2(q_{j-1} - q_j)} \quad j = 2, \ldots, n-1$$
$$\dot{p}_n = \qquad\qquad\qquad 2e^{2(q_{n-1} - q_n)}.$$

We shall assume in addition that $\sum_{j=1}^{n} q_j = \sum_{j=1}^{n} p_j = 0$. The coordinates $q_1, \ldots, q_n$ may be chosen so that this condition is satisfied.

(Strictly speaking, we have defined the *finite, open* Toda lattice; this is also known as the *Toda molecule*. We may also consider the *infinite* Toda lattice, in which there are infinitely many particles, or the *periodic* Toda lattice, in which the first particle interacts with the last.)

Let us now define two $n \times n$ matrices L, M:

$$L = \begin{pmatrix} p_1 & Q_{1,2} & 0 & \cdots & 0 & 0 \\ Q_{1,2} & p_2 & Q_{2,3} & \cdots & 0 & 0 \\ 0 & Q_{2,3} & p_3 & \cdots & 0 & 0 \\ \vdots & \vdots & \vdots & & \vdots & \vdots \\ 0 & 0 & 0 & \cdots & p_{n-1} & Q_{n-1,n} \\ 0 & 0 & 0 & \cdots & Q_{n-1,n} & p_n \end{pmatrix}$$

where $Q_{i,j} = e^{q_i - q_j}$, and

$$M = \begin{pmatrix} 0 & Q_{1,2} & 0 & \cdots & 0 & 0 \\ -Q_{1,2} & 0 & Q_{2,3} & \cdots & 0 & 0 \\ 0 & -Q_{2,3} & 0 & \cdots & 0 & 0 \\ \vdots & \vdots & \vdots & & \vdots & \vdots \\ 0 & 0 & 0 & \cdots & 0 & Q_{n-1,n} \\ 0 & 0 & 0 & \cdots & -Q_{n-1,n} & 0 \end{pmatrix}.$$

By direct calculation, we find the following amazing fact:

Proposition. *The Hamiltonian system for the Toda lattice is equivalent to the Lax equation $\dot{L} = [L, M]$.* ∎

The functions L, M take values in the Lie algebra $\mathbf{sl}_n \mathbf{R}$, and the operation $[L, M] = LM - ML$ is the usual Lie bracket operation.

Let us now consider the decomposition $SL_n\mathbf{R} = SO_n\Delta_n$ of Chapter 3. We have a similar decomposition $SL_n\mathbf{R} = SO_n\Delta_n'$, where Δ_n' denotes the subgroup of lower triangular matrices. Observe that

$$\Delta_n = Z_n A_n N_n, \quad \Delta_n' = Z_n A_n N_n',$$

where Z_n, A_n are (respectively) the subgroups consisting of matrices of the form

$$\begin{pmatrix} \pm 1 & & & \\ & \pm 1 & & \\ & & \ddots & \\ & & & \pm 1 \end{pmatrix}, \begin{pmatrix} d_1 & & & \\ & d_2 & & \\ & & \ddots & \\ & & & d_n \end{pmatrix}$$

with $d_1, \ldots, d_n > 0$, and N_n, N_n' are the subgroups consisting of matrices of the form

$$\begin{pmatrix} 1 & * & * & * \\ & 1 & * & * \\ & & \ddots & \vdots \\ & & & 1 \end{pmatrix}, \begin{pmatrix} 1 & & & \\ * & 1 & & \\ * & * & \ddots & \\ * & * & \cdots & 1 \end{pmatrix}.$$

We shall use the decomposition $G = G_1 G_2$ given by

$$SL_n\mathbf{R} = SO_n \hat{N}_n', \quad \text{where } \hat{N}_n' = A_n N_n'.$$

The corresponding Lie algebras are

$$\mathbf{sl}_n\mathbf{R} = \{X \in M_n\mathbf{R} \mid \text{trace } X = 0\}$$

$$\text{skew}_n\,\mathbf{R} = \{X \in \mathbf{sl}_n\mathbf{R} \mid X^t = -X\}$$

$$\hat{\mathbf{n}}_n' = \{X \in \mathbf{sl}_n\mathbf{R} \mid X = (x_{ij}), x_{ij} = 0 \text{ if } i < j\}$$

$$\mathbf{n}_n' = \{X \in \hat{\mathbf{n}}_n' \mid X = (x_{ij}), x_{11} = \cdots = x_{nn} = 0\}$$

$$\mathbf{a}_n = \{X \in \mathbf{sl}_n\mathbf{R} \mid X = (x_{ij}), x_{ij} = 0 \text{ if } i \neq j\},$$

and the corresponding Lie algebra decomposition is

$$\mathbf{sl}_n\mathbf{R} = \text{skew}_n\,\mathbf{R} \oplus \hat{\mathbf{n}}_n', \quad \text{where } \hat{\mathbf{n}}_n' = \mathbf{a}_n \oplus \mathbf{n}_n'.$$

The resulting decomposition of an element $X \in \mathbf{sl}_n\mathbf{R}$ may be found explicitly, as follows. First, we consider the decomposition

$$\mathbf{sl}_n\mathbf{R} = \mathbf{n}_n \oplus \mathbf{a}_n \oplus \mathbf{n}_n'.$$

For any $X \in \mathbf{sl}_n\mathbf{R}$, we write $X = X_+ + X_0 + X_-$. Hence our decomposition of X is

$$X = (X_+ - X_+^t) + (X_- + X_0 + X_+^t),$$

where $X_+ - X_+^t \in \text{skew}_n\,\mathbf{R}$ and $X_- + X_0 + X_+^t \in \hat{\mathbf{n}}_n'$.

It follows from this explicit formula that $M = \pi_1 L$. Our Hamiltonian system can now be written

$$\dot{L} = [L, \pi_1 L],$$

which is the equation considered (and solved) in section I! Therefore, we arrive at the following conclusion:

Theorem. *The solution of the Hamiltonian system for the Toda lattice is given by* $L(t) = \mathrm{Ad}(\exp tV)_1^{-1}V$, *where* $V = L(0)$. ∎

Let us write down the solution explicitly, in the case $n = 2$. Writing $q_1 = -q$, $q_2 = q$ and $p_1 = -p$, $p_2 = p$, we have

$$L = \begin{pmatrix} p & Q \\ Q & -p \end{pmatrix}, \quad M = \begin{pmatrix} 0 & Q \\ -Q & 0 \end{pmatrix}$$

where $Q = e^{-2q}$. The solution of $\dot{L} = [L, M]$ with

$$L(0) = \begin{pmatrix} 0 & v \\ v & 0 \end{pmatrix}$$

is

$$L(t) = \mathrm{Ad}\left[\exp t \begin{pmatrix} 0 & v \\ v & 0 \end{pmatrix}\right]_1^{-1} \begin{pmatrix} 0 & v \\ v & 0 \end{pmatrix}.$$

In this expression, we have

$$\exp t \begin{pmatrix} 0 & v \\ v & 0 \end{pmatrix} = \begin{pmatrix} \cosh tv & \sinh tv \\ \sinh tv & \cosh tv \end{pmatrix}.$$

The decomposition $SL_2\mathbf{R} = SO_2\hat{N}_2'$ is given by

$$\begin{pmatrix} a & b \\ c & d \end{pmatrix} = \left[\frac{1}{\sqrt{b^2 + d^2}} \begin{pmatrix} d & b \\ -b & d \end{pmatrix}\right]\left[\frac{1}{\sqrt{b^2 + d^2}} \begin{pmatrix} 1 & 0 \\ ab + cd & b^2 + d^2 \end{pmatrix}\right].$$

Hence

$$\left[\exp t \begin{pmatrix} 0 & v \\ v & 0 \end{pmatrix}\right]_1 = \frac{1}{\sqrt{\sinh^2 tv + \cosh^2 tv}} \begin{pmatrix} \cosh tv & \sinh tv \\ -\sinh tv & \cosh tv \end{pmatrix}.$$

We conclude that

$$L(t) = \frac{v}{\sinh^2 tv + \cosh^2 tv} \begin{pmatrix} -2\sinh tv \cosh tv & 1 \\ 1 & 2\sinh tv \cosh tv \end{pmatrix}.$$

This means that

$$p(t) = -v\frac{\sinh 2tv}{\cosh 2tv}, \quad Q(t) = \frac{v}{\cosh 2tv}.$$

The original problem concerns the motion of a single particle on the real line, and the position of the particle is given by $q(t)$, where $Q(t) = e^{-2q(t)}$. It follows that

$$q(t) = -\frac{1}{2}\log\left(\frac{v}{\cosh 2vt}\right)$$

$$= -\frac{1}{2}\log v + \frac{1}{2}\log \cosh 2vt.$$

In Chapter 14, we show how to find such an explicit formula for any n, by using "τ-functions".

Bibliographical comments.

See the comments at the end of Chapter 8.

Chapter 6: Adler-Kostant-Symes

Let us look at the Toda lattice again. We shall try to express everything in terms of the Lie group $SL_n\mathbf{R}$.

1. Lie algebraic coordinates

Let $E_{i,j}$ be the matrix (x_{kl}), where $x_{kl} = \delta_{ik}\delta_{jl}$. We shall need the following elements of the Lie algebra $sl_n\mathbf{R}$:

$$E_j = E_{j,j+1}, \quad 1 \leq j \leq n-1$$
$$E_{-j} = E_{j+1,j}, \quad 1 \leq j \leq n-1$$
$$H_j = E_{j,j} - E_{j+1,j+1}, \quad 1 \leq j \leq n-1.$$

(The vectors E_j, E_{-j} are the "simple root vectors" of $sl_n\mathbf{R}$, and the vectors H_j are the "simple co-root vectors".)

It is easy to verify the following formulae:

$$[H_j, H_k] = 0$$
$$[E_j, E_{-k}] = \delta_{jk}H_k$$
$$[H_j, E_k] = c_{jk}E_k$$

where $C = (c_{ij})$ is the matrix

$$\begin{pmatrix} 2 & -1 & 0 & \cdots & 0 & 0 \\ -1 & 2 & -1 & \cdots & 0 & 0 \\ 0 & -1 & 2 & \cdots & 0 & 0 \\ \vdots & \vdots & \vdots & & \vdots & \vdots \\ 0 & 0 & 0 & \cdots & 2 & -1 \\ 0 & 0 & 0 & \cdots & -1 & 2 \end{pmatrix}.$$

(This matrix is called the "Cartan matrix" of $sl_n\mathbf{R}$.)

If we introduce the variables

$$a_k = e^{q_k - q_{k+1}} = Q_{k,k+1} \quad k = 1, \ldots, n-1$$
$$b_k = \frac{1}{n}[(n-k)(p_1 + \cdots + p_k) - k(p_{k+1} + \cdots + p_n)] \quad k = 1, \ldots, n-1$$

then the matrices L, M of the Toda lattice may be expressed as follows:

$$L = \sum_{j=1}^{n-1} b_j H_j + \sum_{j=1}^{n-1} a_j(E_j + E_{-j})$$

$$M = \sum_{j=1}^{n-1} a_j(E_j - E_{-j}).$$

2. *The Hamiltonian function*

The Lie algebra $sl_n\mathbf{R}$ has a natural Ad-invariant non-degenerate bilinear form

$$(A, B) = \text{trace } AB.$$

(This is – up to a constant multiple – the "Killing form" of $sl_n\mathbf{R}$.) Our Hamiltonian function may be expressed very simply in terms of this form:

$$H(p, q) = \tfrac{1}{2}(L, L).$$

3. *The phase space M*

For the solution of the Hamiltonian system we need the Lie algebra decomposition

$$sl_n\mathbf{R}(= \mathbf{g}) \quad = \quad \text{skew}_n\,\mathbf{R}(= \mathbf{g}_1) \quad \oplus \quad \hat{\mathbf{n}}'_n(= \mathbf{g}_2)$$

(which comes from a decomposition $G = G_1 G_2$ of Lie groups – see Chapter 5). Using the bilinear form, we have an identification

$$(sl_n\mathbf{R})^* \cong sl_n\mathbf{R}.$$

Exercise:

(6.1) Show that (via the above identification) we have

$$(\text{skew}_n\,\mathbf{R})^* \cong (\hat{\mathbf{n}}'_n)^\perp = \mathbf{n}'_n$$
$$(\hat{\mathbf{n}}'_n)^* \cong (\text{skew}_n\,\mathbf{R})^\perp = \text{symm}_n\,\mathbf{R}.$$

(Hint: If (,) is a non-degenerate bilinear form on a vector space $V \oplus W$, then we have an identification $(V \oplus W)^* \to V^* \oplus W^*$ given by $f \mapsto (f|_V, f|_W)$. Next, observe that the image of the natural composition $W^\perp \subseteq V \oplus W \to (V \oplus W)^*$ is contained in the subspace V^* of $(V \oplus W)^*$. Similarly, the image of $V^\perp \subseteq V \oplus W \to (V \oplus W)^*$ is contained in W^*.)

We have already discussed the adjoint action of $SL_n\mathbf{R}$ on $sl_n\mathbf{R}$. It is given by conjugation:

$$\text{Ad}(A)X = AXA^{-1}.$$

We also have the *co-adjoint action* of $SL_n\mathbf{R}$ on $(sl_n\mathbf{R})^*$. It is given by

$$\text{Ad}^*(A)f = g \quad \text{where} \quad g(X) = f(A^{-1}XA).$$

Since we have an Ad-invariant bilinear form on $sl_n\mathbf{R}$, we obtain an equivariant identification $sl_n\mathbf{R} \to (sl_n\mathbf{R})^*$, and the adjoint action on $sl_n\mathbf{R}$ is thereby identified with the co-adjoint action on $(sl_n\mathbf{R})^*$. However, not every[3] Lie algebra **g** admits an Ad-invariant bilinear form, and so the co-adjoint action may differ in an essential way from the adjoint action. This is illustrated by the next example.

Exercise:

(6.2) Show that the co-adjoint action of $\hat{N}'_n$ on $(\hat{\mathbf{n}}'_n)^* \cong (\text{skew}_n \mathbf{R})^\perp = \text{symm}_n \mathbf{R}$ is given explicitly by

$$\text{Ad}^*(A)X = \pi_{symm}(AXA^{-1})$$

where π_{symm} denotes the projection on the first summand of the decomposition $\mathbf{sl}_n\mathbf{R} = \text{symm}_n \mathbf{R} \oplus \mathbf{n}'_n$.

The phase space of our Hamiltonian system may be taken as the subspace of $\text{symm}_n \mathbf{R}$ consisting of matrices of the form

$$\begin{pmatrix}
d_1 & s_1 & 0 & \cdots & 0 & 0 \\
s_1 & d_2 & s_2 & \cdots & 0 & 0 \\
0 & s_2 & d_3 & \cdots & 0 & 0 \\
\vdots & \vdots & \vdots & & \vdots & \vdots \\
0 & 0 & 0 & \cdots & d_{n-1} & s_{n-1} \\
0 & 0 & 0 & \cdots & s_{n-1} & d_n
\end{pmatrix}$$

where $d_i, s_i \in \mathbf{R}, s_i > 0$. In other words, it is the manifold

$$M = \{\sum b_j H_j + \sum a_j(E_j + E_{-j}) \mid a_j, b_j \in \mathbf{R} \text{ and } a_j > 0\}.$$

The Lie algebraic significance of this manifold is explained by the next proposition:

Proposition. *Let* $m = \sum(E_j + E_{-j}) \in \text{symm}_n \mathbf{R}$. *Then* $M = \text{Ad}^*(\hat{N}'_n)m$, *i.e.,* M *is a co-adjoint orbit of the group* $\hat{N}'_n$.

Sketch of the proof. Let $A \in \hat{N}'_n$. Then we have

$$\pi_{symm}(A(\sum E_j + E_{-j})A^{-1}) = \pi_{symm}(A(\sum E_j)A^{-1}).$$

If $X = A(\sum E_j)A^{-1}$, then $\pi_{symm}X = X_+ + X_+^t + X_0$, where we use the notation $X = X_+ + X_0 + X_-$ as in Chapter 5.

---

[3]The Lie algebra of a compact Lie group admits a positive definite Ad-invariant bilinear form. The Lie algebra of a semisimple Lie group (such as $\mathbf{sl}_n\mathbf{R}$) admits a non-degenerate Ad-invariant bilinear form, for example, its Killing form. The group $\hat{N}'_n$ is neither compact nor semisimple, however.

We claim first that $\pi_{symm}(A(\sum E_j)A^{-1}) \in M$. It suffices to show (i) that X is of the form

$$\begin{pmatrix} * & e_1 & 0 & \cdots & 0 & 0 \\ * & * & e_2 & \cdots & 0 & 0 \\ \vdots & \vdots & \vdots & & \vdots & \vdots \\ * & * & * & \cdots & * & e_{n-1} \\ * & * & * & \cdots & * & * \end{pmatrix}$$

and (ii) that $e_1, \ldots, e_{n-1} > 0$. It is easy to verify (i) and (ii).

From the same calculation, one can show that every element of M is of the form $\pi_{symm}(A(\sum E_j)A^{-1})$, for some A. ∎

Exercise:

(6.3) Give a proof of the last proposition for $n = 2$ and $n = 3$.

4. The symplectic structure

Let us recall the main example of Chapter 4, namely the symplectic form ω on an adjoint orbit $M = \mathrm{Ad}(G)P$. For any $X \in \mathbf{g}$, we have a vector field X^* on M, defined by

$$X_R^* = \tfrac{d}{dt}\mathrm{Ad}(\exp tX)R|_0 = [X, R].$$

We define ω by

$$\omega(X_R^*, Y_R^*) = \langle R, [X, Y]\rangle,$$

where $\langle\,,\,\rangle$ denotes an Ad-invariant inner product on $\mathbf{g}$. We have $T_R M_P = \{X_R^* \mid X \in \mathbf{g}\}$, so ω is completely determined by the given formula.

This example is a special case of a more general construction, due to Kirillov, Kostant, and Souriau. Let M be a co-adjoint orbit, i.e., $M = \mathrm{Ad}^*(G)f$, where $f \in \mathbf{g}^*$. For any $X \in \mathbf{g}$, we have a vector field X^* on M, defined this time by

$$X_h^* = \tfrac{d}{dt}\mathrm{Ad}^*(\exp tX)h|_0.$$

Then there is a symplectic form ω on M, defined by

$$\omega(X_h^*, Y_h^*) = h([X, Y]).$$

Exercises:

(6.4) Verify that ω is indeed a symplectic form.

(6.5) Let $(\,,\,)$ be an Ad-invariant non-degenerate bilinear form on $\mathbf{g}$. Show that any co-adjoint orbit may be identified with an adjoint orbit, and that the Kirillov-Kostant-Souriau form ω may be identified with the symplectic form of Chapter 4.

It follows from the last proposition that our phase space

$$M = \text{Ad}^*(\hat{N}'_n)m \subseteq (\hat{\mathfrak{n}}'_n)^*$$

acquires a symplectic form ω. We aim to show that the Hamiltonian system (M, ω, H) is exactly the same as the Toda lattice. This follows from the next proposition:

Proposition. *Let M, ω, H be as above, where*

$$M \subseteq \text{symm}_n \mathbf{R} \subseteq \text{sl}_n \mathbf{R}.$$

Then the Hamiltonian vector field V is given by

$$V_X = [X, \pi_1 X], \quad X \in M,$$

where π_1 is the projection $\text{sl}_n\mathbf{R} = \text{skew}_n\,\mathbf{R} \oplus \hat{\mathfrak{n}}'_n \to \text{skew}_n\,\mathbf{R}$. (Hence, the differential equation of the Hamiltonian system is $\dot{X} = [X, \pi_1 X]$.)

Proof. From the definition of V, we have $dH(Z) = \omega(V_m, Z)$ for all $Z \in T_m M$, $m \in M$. We shall use the embedding $M \subseteq \text{symm}_n\,\mathbf{R}$ given by Exercise (6.2) and the previous proposition, i.e.,

$$M = \{\pi_{symm} A(\textstyle\sum E_j + E_{-j})A^{-1} \mid A \in \hat{N}'_n\}.$$

For any $X \in M$, we have the identification

$$
\begin{aligned}
T_X M &= \{U_X^* \mid U \in \hat{\mathfrak{n}}'_n\} \\
&= \{\tfrac{d}{dt}\pi_{symm}(\exp tU)X(\exp tU)^{-1}|_0 \mid U \in \hat{\mathfrak{n}}'_n\} \\
&= \{\pi_{symm}[U, X] \mid U \in \hat{\mathfrak{n}}'_n\}.
\end{aligned}
$$

We must now interpret the formula $dH(U_X^*) = \omega(V_X, U_X^*)$. The left hand side is:

$$
\begin{aligned}
dH(U_X^*) &= \tfrac{d}{dt}H(\pi_{symm}(\exp tU)X(\exp tU)^{-1}|_0) \\
&= \tfrac{d}{dt}\tfrac{1}{2}(\pi_{symm}(\exp tU)X(\exp tU)^{-1}, \pi_{symm}(\exp tU)X(\exp tU)^{-1})|_0 \\
&= (\pi_{symm}X, \pi_{symm}[U, X]) \\
&= (X, \pi_{symm}[U, X]) \text{ (as } X \in \text{symm}_n\,\mathbf{R}) \\
&= (\pi_2 X, \pi_{symm}[U, X]) \text{ (as } X = \pi_1 X + \pi_2 X, \text{skew}_n\,\mathbf{R} \perp \text{symm}_n\,\mathbf{R}) \\
&= (\pi_2 X, [U, X]) \text{ (using } \hat{\mathfrak{n}}'_n \perp \mathfrak{n}'_n) \\
&= (X, [U, X]) - (\pi_1 X, [U, X]) \\
&= (\pi_1 X, [X, U]) \text{ (as } (X, [U, X]) = 0, \text{ by Exercise (4.2))} \\
(i) \quad &= ([\pi_1 X, X], U) \text{ (by Exercise (4.2)).}
\end{aligned}
$$

For each X, there exists some $W \in \hat{\mathfrak{n}}'_n$ such that $V_X = W^*_X$. Using this, the right hand side is:

$$\omega(W^*_X, U^*_X) = (X, [W, U])$$
$$= ([X, W], U) \text{ (by Exercise (4.2))}$$
$$= -(W^*_X, U)$$
(ii) $$= -(V_X, U).$$

Equations (i) and (ii) hold for all $U \in \hat{\mathfrak{n}}'_n$. In addition, they hold for all $U \in \text{skew}_n \mathbf{R}$, because $V_X, [\pi_1 X, X] \in \text{symm}_n \mathbf{R}$ and $\text{symm}_n \mathbf{R} \perp \text{skew}_n \mathbf{R}$. Since $(\ ,\ )$ is non-degenerate, we deduce that $V_X = -[\pi_1 X, X] = [X, \pi_1 X]$, as required. ∎

Summary: Lie algebraic version of the Toda lattice

(1) The phase space is $M = \text{Ad}^*(\hat{N}'_n)m$, where $m = \sum(E_j + E_{-j})$. We have $M \subseteq (\hat{\mathfrak{n}}'_n)^* \cong (\text{skew}_n \mathbf{R})^{\perp} = \text{symm}_n \mathbf{R}$.

(2) The symplectic form ω is the Kirillov-Kostant-Souriau form.

(3) The Hamiltonian function H is given by $H(X) = \frac{1}{2}(X, X)$.

The differential equation of the Hamiltonian system is $\dot{X} = [X, \pi_1 X]$, where π_1 denotes projection on the first summand of $\mathbf{sl}_n \mathbf{R} = \text{skew}_n \mathbf{R} \oplus \hat{N}'_n$. The solution of this differential equation is $X(t) = \text{Ad}(\exp tV)_1^{-1}V$, where $V = X(0)$.

At this point we may feel that we know all there is to know about the Toda lattice. However, two further questions will be addressed in Chapters 7 and 14:

(1) What is the advantage (if any) of the purely Lie algebraic version of the Toda lattice?

(2) Can the formula for $X(t)$ be made even more explicit? (The original functions $q_1, \ldots, q_n$ cannot yet be read off very easily, if $n > 2$.)

Bibliographical comments.

See the comments at the end of Chapter 8.

Chapter 7: Adler-Kostant-Symes (continued)

I. The generalized Toda lattice.

From Chapter 6, we can see how to generalize the Toda lattice. Let's summarize this briefly, following section 4.5 of Perelomov [1990]. We shall use some terminology and results from Lie theory in this section, which can safely be ignored by the reader who is unconcerned with such abstraction.

The main new ingredient is the real version of the Iwasawa decomposition (see Appendix C of Perelomov [1990], or Helgason [1978]).

Theorem (Iwasawa decomposition). *Let G be a semisimple connected Lie group, which is not compact. Then there is a decomposition*

$$G = KAN, \quad K \cap A = A \cap N = K \cap N = \{e\}$$

where K is compact, A is abelian and N is nilpotent. (In fact, K is a maximal compact subgroup of G.) ∎

We shall need the decompositions

$$G = G_1 G_2 = K\hat{N}, \quad \text{where } \hat{N} = AN$$

and

$$\mathbf{g} = \mathbf{g}_1 \oplus \mathbf{g}_2 = \mathbf{k} \oplus \hat{\mathbf{n}}.$$

Let (,) be the Killing form of $\mathbf{g}$. Using (,), we have an isomorphism $\mathbf{g}^* \cong \mathbf{g}$. Via this isomorphism, the decomposition $\mathbf{g}^* = \mathbf{k}^* \oplus \hat{\mathbf{n}}^*$ corresponds to a decomposition

$$\mathbf{g} = \hat{\mathbf{n}}^{\perp} \oplus \mathbf{k}^{\perp} = \hat{\mathbf{n}}^{\perp} \oplus \mathbf{p}.$$

Let θ be the "Cartan involution" of $\mathbf{g}$, i.e., the automorphism given by multiplication by 1 on $\mathbf{k}$, and multiplication by -1 on $\mathbf{p}$. Let $\{E_\alpha\}_{\alpha \in R^+}$ be a set of positive root vectors (in $\mathbf{g}$), with $\theta(E_\alpha) = -E_{-\alpha}$. Let

$$m = \sum E_\alpha + E_{-\alpha}$$

where the sum is over the simple roots α. Observe that $m \in \mathbf{p}$.

Now we have all the necessary ingredients for the generalized Toda lattice, which is the Hamiltonian system (M, ω, H) defined as follows:

1. The phase space M

$$M = \text{Ad}^*(\hat{N})m$$

(Recall that $\hat{N}$ acts on $\hat{\mathbf{n}}^*$, and that $\hat{\mathbf{n}}^*$ is identified with $\mathbf{p}$.)

2. The symplectic form ω

For any $X, Y \in \mathbf{g}$, and any $h \in \hat{\mathfrak{n}}^*$, $\omega(X_h^*, Y_h^*) = h([X, Y])$.

(This is the Kirillov-Kostant-Souriau symplectic form.)

3. The Hamiltonian function H

$H(X) = \frac{1}{2}(X, X)$.

The Hamiltonian system of Chapter 6 (the Toda lattice) is evidently a special case of the above definition. By the same argument as in Chapter 6, we have:

Theorem. *The differential equation of the above Hamiltonian system is* $\dot{X} = [X, \pi_1 X]$, *where* π_1 *denotes projection on the first summand of* $\mathbf{g} = \mathbf{k} \oplus \hat{\mathfrak{n}}$. *The solution of this differential equation is* $X(t) = \mathrm{Ad}(\exp tV)_1^{-1} V$, *where* $V = X(0)$. ∎

It is possible to interpret the generalized Toda lattice as a system of particles, where the forces between the particles depend on the (Dynkin diagram of the) Lie group G.

II. Integrals of motion, commuting Hamiltonians.

Two questions arise from the previous discussion:

(1) What is the significance of the point m? (Could we could use *any* m in **p**?)

(2) What is the significance of the function $H(X) = \frac{1}{2}(X, X)$? (Could we use *other* functions?)

We shall now consider these questions.

An important idea in the theory of Hamiltonian systems is the idea of "integrals of motion" or "conserved quantities". An *integral* of a Hamiltonian system (M, ω, H) is a function $I : M \to \mathbf{R}$ such that $I(X(t))$ is constant, for any solution X of the differential equation. If V_I is the Hamiltonian vector field associated to I, then we have:

$$\tfrac{d}{dt} I(X(t)) = dI(\dot{X}) = dI(V) = \omega(V_I, V).$$

It is obvious that the function H itself is an integral, because $V = V_H$. (This corresponds to "conservation of energy" in mechanics.) If I is an integral, then (under suitable conditions) we can obtain a new Hamiltonian system on a smaller phase space

$$M' = I^{-1}(c)/\mathbf{R}, \quad \dim M' = \dim M - 2.$$

(The action of **R** on $I^{-1}(c)$ is given by the "flow" of the vector field V_I.) The new system is a "reduction" of the old system.

We shall not explain the details of this procedure, which is called *symplectic reduction* (see Arnold [1978]; Perelomov [1990]). However, we remark that we have already seen an example of this procedure, for the Toda lattice. Namely, if we impose the condition $\sum q_i = 0$, it follows that $\sum p_i = 0$, and so we reduce the phase space from $\mathbf{R}^{2n}$ to $\mathbf{R}^{2n-2}$. (This corresponds to "conservation of momentum".)

If $I_1, \ldots, I_m$ are independent integrals, and if they are in involution, then (under suitable conditions) we obtain a new Hamiltonian system on a phase space M', with $\dim M' = \dim M - 2m$. (We say that $I_1, \ldots, I_m$ are *independent* if $dI_1, \ldots, dI_m$ are linearly independent at each point, and that $I_1, \ldots, I_m$ are *in involution* if $\omega(V_i, V_j) = 0$ for $1 \leq i, j \leq m$, where V_i is the Hamiltonian vector field associated to I_i.)

In general, integrals (apart from H) do not exist (and even if they exist, they are not easy to find). If there exist n independent integrals in involution, where $\dim M = 2n$, then we say that the Hamiltonian system (M, ω, H) is *completely integrable*. This is a very special situation. For further explanation of this, and for many examples from mechanics, we refer again to Arnold [1978]; Perelomov [1990].

It turns out that the Toda lattice (and the generalized Toda lattice) is completely integrable! This important theoretical property provides an answer to questions (1) and (2) above. Therefore, we shall now describe some integrals of the (generalized) Toda lattice.

We say that a function $f : \mathbf{g} \to \mathbf{R}$ is *invariant* if it is invariant under the adjoint action, i.e.,

$$f(\operatorname{Ad} gX) = f(X) \quad \text{for all } g \in G, X \in \mathbf{g}.$$

Lemma 1. *Let $f : \mathbf{g} \to \mathbf{R}$ be an invariant function. Then:*

(1) $(df)_X[X, Y] = 0$ for all $X, Y \in \mathbf{g}$, and

(2) $(df)_{\operatorname{Ad} gX} \operatorname{Ad} gY = (df)_X(Y)$ for all $X, Y \in \mathbf{g}$. ∎

Exercises

(7.1) Prove Lemma 1. (Differentiate the equation $f(\operatorname{Ad} gX) = f(X)$. For (1), put $g = \exp tY$. For (2) replace X by $X + tY$.)

(7.2) Let ∇f be the gradient of f with respect to $(\ ,\ )$. Using Lemma 1, show that (1) $[(\nabla f)_X, X] = 0$ for all $X \in \mathbf{g}$, and (2) $\operatorname{Ad} g(\nabla f)_X = (\nabla f)_{\operatorname{Ad} gX}$ for all $X \in \mathbf{g}, g \in G$. (Here, f is an invariant function on $\mathbf{g}$.)

Using this lemma, we obtain the following theorem:

Theorem. *Let $f_1, \ldots, f_m : \mathbf{g} \to \mathbf{R}$ be invariant functions. Then*

(1) $f_1, \ldots, f_m$ are integrals of the generalized Toda lattice.

(2) $\omega(V_i, V_j) = 0$ for $1 \leq i, j \leq m$, where V_i denotes the Hamiltonian vector field associated to f_i.

Proof. (1) Let X be a solution of the generalized Toda lattice, i.e., a solution of the differential equation $\dot{X} = [X, \pi_1 X]$. Let $f : \mathbf{g} \to \mathbf{R}$ be an invariant function. Then

$$\tfrac{d}{dt} f(X(t)) = (df)_{X(t)} \dot{X}(t) = (df)_{X(t)} [X(t), \pi_1 X(t)]$$

and this is zero by Lemma 1. (2) Let $f, g : \mathbf{g} \to \mathbf{R}$ be invariant functions, and let V_f, V_g be the corresponding Hamiltonian vector fields. We shall prove later (in Lemma 2) that $(V_g)_X = [X, \pi_1(\nabla g)_X]$. Hence

$$\omega((V_f)_X, (V_g)_X) = (df)_X((V_g)_X) = (df)_X([X, \pi_1(\nabla g)_X])$$

and again this is zero by Lemma 1. ∎

For the (original) Toda lattice, we have $n - 1$ invariant functions:

$$f_i : \mathbf{sl}_n \mathbf{R} \to \mathbf{R}, \quad f_i(X) = \text{trace } X^i, \quad i = 2, \ldots, n.$$

Since dim $M = 2n - 2$, we see that the Toda lattice is completely integrable. Thus, we can answer question (1): The importance of the point m is that its co-adjoint orbit is "small". If we choose *any* point of symm$_n$ **R**, then we obtain a Hamiltonian system, but this Hamiltonian system will not in general be completely integrable. A similar statement applies to the generalized Toda lattice.

It is easy to see directly that the functions $X \mapsto \text{trace } X^i$ are integrals of the Toda lattice. We know that the explicit solution is given by $X(t) = \text{Ad}(\exp tV)_1^{-1} V$, so the eigenvalues of the matrix $X(t)$ are constant (independent of t). Hence the symmetric functions of the eigenvalues are constant, and hence the functions $f_i(X(t)) = \text{trace } X(t)^i$ are constant. This suggests a generalization of the Toda lattice, which is based on the following result:

Lemma 2. *Let $f : \mathbf{g} \to \mathbf{R}$ be an invariant function. Then the associated Hamiltonian vector field V_f is given by*

$$(V_f)_X = [X, \pi_1(\nabla f)_X]$$

where ∇f is the gradient of f with respect to the bilinear form $(\ ,\ )$.

Proof. In the case $f = H$, we have already proved this in Chapter 6. The proof in the general case is very similar; it depends on the following calculation of $df(U_X^*)$, for $U \in \mathbf{p}$:

$$
\begin{aligned}
(df)_X(U_X^*) &= ((\nabla f)_X, \pi_\mathbf{p}[U, X]) \quad \text{by definition of } \nabla f \\
&= (\pi_2(\nabla f)_X, \pi_\mathbf{p}[U, X]) \\
&= (\pi_2(\nabla f)_X, [U, X]) \\
&= ((\nabla f)_X, [U, X]) - (\pi_1(\nabla f)_X, [U, X]) \\
&= -(\pi_1(\nabla f)_X, [U, X])
\end{aligned}
$$

where the last line follows from Lemma 1. ∎

Hence, we have a new Hamiltonian system (M, ω, f) for every invariant function f! The differential equation of this Hamiltonian system is

$$\dot{X} = [X, \pi_1(\nabla f)_X].$$

This is very similar to the Lax equations from Chapter 5. In fact, we can solve the equation explicitly by the same method that we used in Chapter 5:

Theorem. *The solution of the differential equation*

$$\dot{X} = [X, \pi_1(\nabla f)_X], \quad X(0) = V$$

is given by $X(t) = \mathrm{Ad}(\exp t(\nabla f)_V)_1^{-1} V$.

Proof. Let $g(t) = \exp tW$ and $u(t) = g(t)_1$, with $W = (\nabla f)_V$. As in Chapter 5, we see that

(a) $X = \mathrm{Ad}\, u^{-1} V$ is a solution of the equation $\dot{X} = [X, u^{-1}\dot{u}]$, and

(b) $u^{-1}\dot{u} = \pi_1 g_1^{-1} W g_1$.

By Exercise (7.2), we have $g_1^{-1} W g_1 = g_1^{-1}(\nabla f)_V g_1 = (\nabla f)_{g_1^{-1} V g_1} = (\nabla f)_X$. ∎

This gives an answer to question (2).

It is well known that

$$I(\mathbf{g}) = \{f : \mathbf{g} \to \mathbf{R} \mid f \text{ is invariant}\}$$

is a finitely generated algebra. Let $f_1, \ldots, f_m$ be a set of independent generators. (In the case of $\mathbf{g} = \mathrm{sl}_n\mathbf{R}$, we have the functions $f_i(X) = \mathrm{trace}\, X^i$, $i = 2, \ldots, n$.) Therefore, we have a natural "hierarchy" of Hamiltonian systems, namely (M, ω, f_i), $i = 1, \ldots, m$. This "Toda hierarchy" has the following properties:

(1) The differential equation of the i-th Hamiltonian system (M, ω, f_i) is $\dot{X} = [X, \pi_1(\nabla f_i)_X]$.

(2) The explicit solution of this equation is $X(t) = \mathrm{Ad}(\exp t(\nabla f_i)_V)_1^{-1} V$, where $V = X(0)$.

(3) The functions $f_1, \ldots, \hat{f}_i, \ldots, f_m$ (with f_i omitted) are integrals of the i-th Hamiltonian system.

(We have already proved (1) and (2). To prove (3), we must prove that $\omega(V_i, V_j) = 0$. We have $\omega((V_i)_X, (V_j)_X) = (df_i)_X(V_j)_X$. By Lemma 2 this is $(df_i)_X([X, \pi_1(\nabla f_j)_X]$, which is zero by Lemma 1. It should also be noted that $df_1, \ldots, df_m$ are independent; this follows from Euler's formula for the derivative of a homogeneous polynomial.)

Remark: If $f, g : \mathbf{g} \to \mathbf{R}$ are (not necessarily invariant) functions, we define the *Poisson bracket* of f and g to be

$$\{f, g\} = \omega(V_f, V_g).$$

If $\{f, g\} = 0$, we say that f and g *Poisson commute*. Thus, the invariant functions $f_1, \ldots, f_m$ are sometimes called "commuting Hamiltonians". The concept of a Poisson structure is more general than the concept of a symplectic structure. Further information on this subject may be found in Perelomov [1990].

Bibliographical comments.

See the comments at the end of Chapter 8.

Chapter 8: Concluding remarks
on one-dimensional Lax equations

I. Further generalizations of the Toda lattice.

In Chapter 7, we used the Iwasawa decomposition $G = KAN = K\hat{N}$ of a (real) Lie group to obtain many examples (M, ω, f) of Hamiltonian systems. Namely, we took

(1) $M = \text{Ad}^*(\hat{N})m$, for some $m \in \mathbf{p}$,

(2) $\omega =$ the Kirillov-Kostant-Souriau symplectic form, and

(3) $f : \mathbf{g} \to \mathbf{R}$ an invariant function.

It is not surprising that this idea can be generalized even further. In this section we discuss three such generalizations. For more details we refer to section §1.12 of Perelomov [1990].

Modification of the group decomposition

Let G be a Lie group, with Lie algebra $\mathbf{g}$, and let $\mathbf{g}_1, \mathbf{g}_2$ be subalgebras of $\mathbf{g}$ such that $\mathbf{g} = \mathbf{g}_1 \oplus \mathbf{g}_2$. Let G_1, G_2 be the corresponding subgroups of G. *(Warning: G is not necessarily equal to $G_1 G_2$!)* Let $(\ ,\ )$ be a non-degenerate bilinear form on $\mathbf{g}$. Then we have

$$\mathbf{g} = \mathbf{g}_2{}^\perp \oplus \mathbf{g}_1{}^\perp \quad (\text{from } \mathbf{g} \cong \mathbf{g}^* \cong \mathbf{g}_1{}^* \oplus \mathbf{g}_2{}^*)$$

and we can define

$$M = \text{Ad}^*(G_2)m \quad (\text{for } m \in \mathbf{g}_1{}^\perp).$$

We have the usual Kirillov-Kostant-Souriau symplectic form ω on M. If $f : \mathbf{g} \to \mathbf{R}$ is an invariant function, then we obtain a Hamiltonian system (M, ω, f). The differential equation of this system is

$$\dot{X} = [X, \pi_1(\nabla f)_X]$$

(by the usual calculation).

If $G \neq G_1 G_2$, in general we do *not* have the usual formula for the solution $X : \mathbf{R} \to M$ of the above differential equation. However, if we apply the Inverse Function Theorem to the map $G_1 \times G_2 \to G$, $(g, h) \mapsto gh$, then we see *locally* that $G = G_1 G_2$ (near e). So we can show that there is a local solution

$$X(t) = \text{Ad}(\exp t(\nabla f)_V)_1^{-1} V, \quad t \in (-\epsilon, \epsilon)$$

for some $\epsilon > 0$.

Example:

Let $G = SL_n\mathbf{R}$. We have the decomposition

$$\mathbf{g} = \mathbf{g}_1 \oplus \mathbf{g}_2 = \hat{\mathbf{n}}'_n \oplus \mathbf{n}_n$$

("lower triangular" and "strictly upper triangular" matrices, respectively). Here, $SL_n\mathbf{R} \neq \hat{N}'_n N_n$.

Exercise:

(8.1) Verify that $SL_2\mathbf{R} \neq \hat{N}'_2 N_2$.

The decomposition $\mathbf{sl}_n\mathbf{R} = \hat{\mathbf{n}}'_n \oplus \mathbf{n}_n$ and the corresponding "partial decomposition" $SL_n\mathbf{R} \supseteq \hat{N}'_n N_n$ are called *Gauss decompositions*. (This name is explained in Chapter 14.)

Modification of the phase space

As a second generalization, let us consider the Lax equation

$$\dot{X} = [X, \pi_1 X] \quad \text{for } X : \mathbf{R} \to \mathbf{g}$$

(or more generally $\dot{X} = [X, \pi_1(\nabla f)_X]$). Here we assume that $\mathbf{g} = \mathbf{g}_1 \oplus \mathbf{g}_2$, but we do *not* choose a co-adjoint orbit. The argument of Chapter 5 (and the previous paragraph) shows that we have a local solution $X(t) = \text{Ad}(\exp tV)_1^{-1} V$, $t \in (-\epsilon, \epsilon)$, for some $\epsilon > 0$. We shall discuss in detail an example of this situation.

Example:

Let $G = SL_n\mathbf{R}$, and consider the Gauss decomposition $\mathbf{g} = \hat{\mathbf{n}}'_n \oplus \mathbf{n}_n$ (as in the previous example). Consider the Lax equation

$$\dot{X} = [X, \pi_{\mathbf{n}_n} X]$$

where X has the following special form:

$$X = \begin{pmatrix} b_1 & a_1 & 0 & \cdots & 0 & 0 \\ 1 & b_2 & a_2 & \cdots & 0 & 0 \\ 0 & 1 & b_3 & \cdots & 0 & 0 \\ \vdots & \vdots & \vdots & & \vdots & \vdots \\ 0 & 0 & 0 & \cdots & b_{n-1} & a_{n-1} \\ 0 & 0 & 0 & \cdots & 1 & b_n \end{pmatrix}.$$

(Note that the Lax equation becomes $\dot{Y} = [Y, \pi_{\hat{\mathbf{n}}'_n} Y]$, if we put $Y = -X$.) We have

$$\pi_{\mathbf{n}_n} X = \begin{pmatrix} 0 & a_1 & 0 & \cdots & 0 & 0 \\ 0 & 0 & a_2 & \cdots & 0 & 0 \\ 0 & 0 & 0 & \cdots & 0 & 0 \\ \vdots & \vdots & \vdots & & \vdots & \vdots \\ 0 & 0 & 0 & \cdots & 0 & a_{n-1} \\ 0 & 0 & 0 & \cdots & 0 & 0 \end{pmatrix},$$

and so the equation reduces to the following system:

$$\dot{a}_i = a_i(b_i - b_{i+1})$$
$$\dot{b}_i = a_{i-1} - a_i.$$

If we make the change of variable

$$a_i = 4\alpha_i^2, \quad b_i = 2\beta_i$$

we obtain the system

$$\dot{\alpha}_i = \alpha_i(\beta_i - \beta_{i+1})$$
$$\dot{\beta}_i = 2(\alpha_{i-1}^2 - \alpha_i^2).$$

This is the same as the Toda lattice system! To see this, recall that the Toda lattice system is

$$\dot{q}_i = p_i$$
$$\dot{p}_i = 2e^{2(q_{i-1}-q_i)} - 2e^{2(q_i-q_{i+1})}.$$

If we put $Q_i = e^{q_i - q_{i+1}}$ then we obtain the system

$$\dot{Q}_i = Q_i(p_i - p_{i+1})$$
$$\dot{p}_i = 2(Q_{i-1}^2 - Q_i^2).$$

This is the same as the earlier system. (So, we see that *different* Lax equations may be equivalent to the *same* differential equation.) We know from our earlier discussion of the Toda lattice that the solution is defined for all t (i.e., in this case, $\epsilon = \infty$).

Let us verify that this agrees with the method of Chapter 5, in the case $n = 2$. We have

$$X = \begin{pmatrix} b & a \\ 1 & -b \end{pmatrix}, \quad \pi_{n_2} X = \begin{pmatrix} 0 & a \\ 0 & 0 \end{pmatrix}.$$

The solution of $\dot{X} = [X, \pi_{n_2} X]$ with

$$X(0) = \begin{pmatrix} 0 & 1 \\ 1 & 0 \end{pmatrix}$$

is

$$X(t) = \mathrm{Ad}\left[\exp t \begin{pmatrix} 0 & 1 \\ 1 & 0 \end{pmatrix}\right]_1^{-1} \begin{pmatrix} 0 & 1 \\ 1 & 0 \end{pmatrix}.$$

We have
$$\exp t \begin{pmatrix} 0 & 1 \\ 1 & 0 \end{pmatrix} = \begin{pmatrix} \cosh t & \sinh t \\ \sinh t & \cosh t \end{pmatrix}.$$

The Gauss factorization is given by
$$\begin{pmatrix} a & b \\ c & d \end{pmatrix} = \begin{pmatrix} 1 & b/d \\ 0 & 1 \end{pmatrix} \begin{pmatrix} 1/d & 0 \\ c & d \end{pmatrix}.$$

(This is valid only when $d > 0$!) We obtain
$$X(t) = \begin{pmatrix} -\tanh t & 1 - \tanh^2 t \\ 1 & \tanh t \end{pmatrix}.$$

Thus
$$a(t) = \frac{1}{\cosh^2 t}, \quad b(t) = -\frac{\sinh t}{\cosh t},$$

i.e.,
$$\alpha(t) = \frac{1}{2\cosh t}, \quad \beta(t) = -\frac{\sinh t}{2\cosh t}.$$

This agrees with the explicit formula obtained in Chapter 5, if we take $v = \frac{1}{2}$ in that formula.

Modification of the Lax equation

The form of the Lax equation itself can be generalized in various ways. We shall discuss two of these.

The analysis of the Lax equation in Chapter 7 depended only on certain properties of the vector field ∇f, where $f : \mathbf{g} \to \mathbf{R}$ is an invariant function. More generally, we say that a function $J : \mathbf{g} \to \mathbf{g}$ is an *invariant vector field* if $\mathrm{Ad}(g)J_Y = J_{\mathrm{Ad}(g)Y}$ for all $g \in G$, $Y \in \mathbf{g}$. (Because we want to think of J as a vector field on $\mathbf{g}$, we use the notation J_Y for the value of J at Y, instead of $J(Y)$.) The various properties of ∇f used in Chapter 7 are special cases of the following general results:

Lemma.

(1) If J is an invariant vector field, then $[X, J_Y] = ([X,Y] \cdot J)_Y$ and $[X, J_X] = 0$ (for any $X, Y \in \mathbf{g}$).

(2) If J_1, J_2 are invariant vector fields, then $[(J_1)_Y, (J_2)_Y] = 0$ (for any $Y \in \mathbf{g}$), i.e., the pointwise Lie bracket of J_1, J_2 is zero.

Proof. (1) (By $[X,Y] \cdot J$ we mean the result of differentiating the function J in the direction of the vector $[X,Y]$.) The first equation follows from the definition of an invariant vector field, on putting $g = \exp tX$ and then differentiating with respect to t. The second equation is obtained by putting $Y = X$ in the first equation. (2) By (1) we have $[(J_1)_Y, (J_2)_Y] = ([J_1)_Y, Y] \cdot J_2)_Y = 0$. ∎

Exactly as in section II of Chapter 7, we obtain:

Theorem. *Let $J : \mathbf{g} \to \mathbf{g}$ be an invariant vector field. The solution of the differential equation*

$$\dot{X} = [X, \pi_1 J_X], \quad X(0) = V$$

is given by $X(t) = \mathrm{Ad}(\exp tJ_V)_1^{-1}V$. ∎

As a final generalization, we mention Lax equations which arise from "R-matrices". Let $\mathbf{g}$ be a Lie algebra, and let $R : \mathbf{g} \to \mathbf{g}$ be a linear transformation. For any $X, Y \in \mathbf{g}$, we define

$$[X, Y]_R = [RX, Y] + [X, RY].$$

We say that R is a *classical R-matrix* if the bilinear form $[\ ,\ ]_R$ defines a Lie algebra structure on $\mathbf{g}$. (In this case, $\mathbf{g}$ possesses two Lie algebra structures.)

Example:

If $\mathbf{g} = \mathbf{g}_1 \oplus \mathbf{g}_2$, we may define $R : \mathbf{g} \to \mathbf{g}$ by $R(X) = \frac{1}{2}(\pi_1 X - \pi_2 X)$, where π_1, π_2 are the projection maps to $\mathbf{g}_1, \mathbf{g}_2$.

Exercise:

(8.2) Show that R (in the last example) is a classical R-matrix. (The main problem is to verify the Jacobi identity.)

We may consider the Lax equation $\dot{X} = [X, R(\nabla f)_X]$ for any invariant function $f : \mathbf{g} \to \mathbf{R}$. For the above example, this equation reduces to the usual equation $\dot{X} = [X, \pi_1(\nabla f)_X]$. More generally still, we may consider the Lax equation $\dot{X} = [X, RJ_X]$, for any invariant vector field J.

II. Group theory.

We shall describe here some elementary group theoretic aspects of Lax equations. Although these give no new information in our simple examples, they are important in the study of more complicated integrable systems. There are two ways in which group actions arise: (a) each solution of the Lax equation is given by the action of a one parameter subgroup, and (b) there is a group action on the space of solutions.

Regarding (a), recall that we have considered two basic examples of Lax equations so far. They are

(i) $\dot{X} = [X, \pi_1 W], \quad X(0) = V$
with solution $X(t) = \mathrm{Ad}(\exp -tW)_1 V$),

(ii) $\dot{X} = [X, \pi_1(\nabla f)_X], \quad X(0) = V$
with solution $X(t) = \mathrm{Ad}(\exp t(\nabla f)_V)_1^{-1}V)$.

In each case, the solution is given by the action of a one parameter subgroup $\{\exp tU \mid t \in \mathbf{R}\}$ on the initial point V. It turns out that the two actions arise naturally, and are closely related.

To explain this, we need the following algebraic observation. Let G be (any) group. Assume that $G = G_1 G_2$, where G_1, G_2 are subgroups of G, with $G_1 \cap G_2 = \{e\}$. Then we obtain two group actions of (the group) G on (the set) G_1, from the formulae

(1) $$g \cdot h = (gh)_1$$

(2) $$g \cdot h = (gh^{-1})_1^{-1}.$$

(As usual, we write $g = g_1 g_2$ for the factorization of $g \in G$.) These two actions are called "dressing actions".

Exercise:

(8.3) Verify that (1) and (2) define group actions, i.e., that $ab \cdot c = a \cdot (b \cdot c)$ for any $a, b \in G$, $c \in G_1$.

The solution of (i) is given by $\mathrm{Ad}(\exp -tW \cdot e)V$, where $\exp -tW \cdot e$ is defined using (1); the solution of (ii) is given by $\mathrm{Ad}(\exp t(\nabla f)_V \cdot e)V$, where $\exp t(\nabla f)_V \cdot e$ is defined using (2).

If $g \in G_2$ and $h \in G_1$, then the first action $g \cdot h = (gh)_1$ has a particularly simple interpretation: It is a measure of the extent to which G_1 and G_2 commute.

Regarding (b), let us consider a solution $X = \mathrm{Ad}\, u^{-1}V$, where $u(t) = (\exp t(\nabla f)_V)_1$. Let g be any element of G_2. We compute $g \cdot u$:

$$\begin{aligned}
g \cdot u(t) &= (g(\exp t(\nabla f)_V)_1)_1 \\
&= (g(\exp t(\nabla f)_V)_1(\exp t(\nabla f)_V)_2))_1 \\
&= (g\exp t(\nabla f)_V)_1 \\
&= (g(\exp t(\nabla f)_V)g^{-1}g)_1 \\
&= (g(\exp t(\nabla f)_V)g^{-1})_1 \quad \text{(since } g \in G_2) \\
&= (\exp tg(\nabla f)_V g^{-1})_1 \\
&= (\exp t\nabla f_{gVg^{-1}})_1.
\end{aligned}$$

Thus, $g \cdot u$ has the same form as u, but with gVg^{-1} instead of V. We conclude that the action of G_2 on G_1 induces an action of G_2 on solutions of the Lax equation.

III. The Toda flow and the gradient flow.

In this section we describe a relationship between the Lax equations (i) and (ii) above. This was discovered only recently (see Bloch *et al.* [1992]).

The Toda lattice (equation (ii)) is defined in terms of the co-adjoint orbit $\mathrm{Ad}^*(\hat{N})'m$. On the other hand, we know that the explicit solution of the Toda lattice lies in the adjoint orbit $\mathrm{Ad}(K)m$. Therefore, we ask: Is it possible to define the Toda lattice in terms of $\mathrm{Ad}(K)m$? It turns out that the answer to this question is "yes". Moreover, it turns out that there is a surprising relationship between equations (i) and (ii) in this case.

To explain this, we shall begin by reviewing the example of equation (i) which was given in Chapter 4. Let G be a compact Lie group, and let $P, Q \in \mathbf{g}$. Then we have a Hamiltonian system (M_P, ω, H^Q). The definitions are as follows:

$M_P = \mathrm{Ad}(G)P$

$\omega(X_R^*, Y_R^*) = (R, [X, Y])$ (where $(\,,\,)$ is an Ad-invariant inner product on $\mathbf{g}$, and where $X_R^* = [X, R] \in T_R M_P$, for any $R \in M_P \subseteq \mathbf{g}$).

$H^Q(R) = (R, Q)$.

The differential equation of this Hamiltonian system is $\dot{X} = [X, Q]$, and the solution is $X(t) = \mathrm{Ad}(\exp -tQ)V$ (where $V = X(0)$).

We have another differential equation here, namely the equation for the "gradient flow" of H^Q. This equation is:

$$\dot{X} = (\nabla H^Q)_X = JQ_X^* = J[Q, X].$$

This is not a Lax equation. However, its solution is given by the explicit formula $X(t) = \mathrm{Ad}(\exp \sqrt{-1}\,tQ)_u V$. This formula is similar to the formula for the solution of the Lax equation (ii).

Now we come to the main point of this section: If we *modify* the gradient flow equation, *then* we get a Lax equation! The gradient of H^Q (i.e., ∇H^Q) depends on the choice of a Riemannian metric $\langle\,,\,\rangle$ on M_P. So far, we have used the Riemannian metric $\langle\,,\,\rangle$ such that

$$\omega(X_R^*, Y_R^*) = \langle X_R^*, JY_R^* \rangle$$

(where J is a certain complex structure on M_P). From now on we shall use a different Riemannian metric $\langle\,,\,\rangle'$. This is defined by the formula

$$\langle X_R^*, Y_R^* \rangle' = (X_1, Y_1)$$

where $X = X_1 + X_2$ denotes the expression of $X \in \mathbf{g}$ with respect to the decomposition $\mathbf{g} = \mathrm{Im}\, A_R \oplus \mathrm{Ker}\, A_P$.

Exercise:

(8.4) Show that $\langle\,,\,\rangle'$ is well defined.

Proposition. *The gradient of H^Q with respect to the Riemannian metric $\langle\,,\,\rangle'$ is given by $(\nabla H^Q)_R = [R, [Q, R]]$.*

Proof. By definition of ∇H^Q, we have (for all $X \in \mathbf{g}$)

$$
\begin{aligned}
\langle(\nabla H^Q)_R, X_R^*\rangle' &= (dH^Q)_R(X_R^*) \\
&= \tfrac{d}{dt} H^Q(\mathrm{Ad}(\exp tX)R)|_0 \\
&= ([X, R], Q) \\
&= (X, [R, Q]).
\end{aligned}
$$

On the other hand, by definition of $\langle\,,\,\rangle'$, we have

$$
\begin{aligned}
\langle(\nabla H^Q)_R, X_R^*\rangle' &= (Y_1, X_1) \quad \text{where we write } (\nabla H^Q)_R = Y_R^* \\
&= (Y_1, X) \quad \text{as Ker } A_R \perp \mathrm{Im}\, A_R \\
&= (Y - Y_2, X).
\end{aligned}
$$

Hence $[R, Q] = Y - Y_2$, and on applying $[R,\ ]$ we obtain $[R, [R, Q] = [R, Y] = -Y_R^* = \nabla H^Q)_R$, as required. ∎

Thus, the new gradient flow equation is

$$
\dot{X} = [X, [Q, X]],
$$

which is a Lax equation.

Let us consider this equation for $G = SL_n\mathbf{R}$. (This G is not compact, but the above discussion is still valid if we take $(\,,\,)$ to be the Ad-invariant bilinear form $(A, B) = \mathrm{trace}\, AB$. The only difference is that $\langle\,,\,\rangle'$ is not necessarily positive definite in this case.) We take

$$
Q = \begin{pmatrix} n & & & \\ & n-1 & & \\ & & \ddots & \\ & & & 1 \end{pmatrix}.
$$

If we take

$$
X = \begin{pmatrix}
p_1 & Q_{1,2} & 0 & \cdots & 0 & 0 \\
Q_{1,2} & p_2 & Q_{2,3} & \cdots & 0 & 0 \\
0 & Q_{2,3} & p_3 & \cdots & 0 & 0 \\
\vdots & \vdots & \vdots & & \vdots & \vdots \\
0 & 0 & 0 & \cdots & p_{n-1} & Q_{n-1,n} \\
0 & 0 & 0 & \cdots & Q_{n-1,n} & p_n
\end{pmatrix}
$$

then we have

$$
[Q,X] = \begin{pmatrix}
0 & Q_{1,2} & 0 & \cdots & 0 & 0 \\
-Q_{1,2} & 0 & Q_{2,3} & \cdots & 0 & 0 \\
0 & -Q_{2,3} & 0 & \cdots & 0 & 0 \\
\vdots & \vdots & \vdots & & \vdots & \vdots \\
0 & 0 & 0 & \cdots & 0 & Q_{n-1,n} \\
0 & 0 & 0 & \cdots & -Q_{n-1,n} & 0
\end{pmatrix}.
$$

Hence, the equation $\dot{X} = [X,[Q,X]]$ is the equation of the Toda lattice! This shows that the "Toda flow" can be identified with a "gradient flow".

Bibliographical comments for Chapters 5-8.

Lax equations are ubiquitous in the modern theory of integrable systems. The Lax form of the Toda lattice was discovered by Flaschka, and the exponential formula for the solution was obtained by Symes. The co-adjoint orbit point of view is due to Adler and Kostant. (The title "Adler-Kostant-Symes" is taken from Burstall and Pedit [1994], which contains a summary of this material.) A general treatment of both the non-periodic and periodic Toda lattices is given in Goodman and Wallach [1984b]. This article also has references to the work of the above authors.

We have mainly followed the exposition of the book by Perelomov [1990], which is an excellent survey of integrable systems from the point of view of Lie groups and Lie algebras. The reader will find a much more comprehensive treatment in Perelomov [1990]; Arnold and Novikov [1990; 1994], as well as many references to the parallel Russian literature.

Atiyah [1979] contains several accessible articles which are related to the material of the preceding chapters. There is a survey of Lie algebras by MacDonald, a survey of compact Lie groups (including height functions on adjoint orbits, as discussed in section II of Chapter 4) by Bott, and an article on the Toda lattice by Kostant.

The idea of "integrals of motion" is at the heart of the relationship between integrable systems and group theory. The most famous classical result is Noether's theorem, which says (roughly) that integrals of motion correspond to symmetries of the system. In modern language, this can be expressed as follows. Assume that a group G acts on a phase space M of a Hamiltonian system, preserving the symplectic form ω. Then (under mild conditions) we have a map $\mu : M \to \mathbf{g}^*$ such that $d\mu_m(V_m)(X) = \omega(V_m, X_m^*)$ for any vector field V on M and any $X \in \mathbf{g}$. This map μ is called the "moment map" or "momentum map"; it is a vector-valued map whose components are the integrals of motion which arise from G. Under further conditions, the space $\mu^{-1}(c)/G$ (for certain $c \in \mathbf{g}^*$) inherits a symplectic form.

For a completely integrable system, one has such an action of $G = \mathbf{R}^k$, with $k = \frac{1}{2} \dim M$. Each level set $\mu^{-1}(c)$ has a linear structure, and in favourable situations this leads to "action-angle coordinates" on M, in terms of which the differential equation may be solved "by quadrature". This is (essentially) the content of Liouville's theorem (see Arnold [1978]). (We postpone discussion of this kind of explicit integration of the Toda lattice until Chapter 14.)

The importance of completely integrable systems comes from this possibility of explicit integration. It is in general a hopeless task to find an explicit formula for the integral curves of a vector field, but the special case of a completely integrable system is considerably more tractable. Remarkably, many examples from classical mechanics are of this type.

The concept of an R-matrix is another prominent feature of the theory of integrable systems. As this has no immediate application to our examples, we mention it only briefly, in Chapter 8. The reader will find further information in Perelomov [1990].

The idea of a "dressing action" comes originally from a part of the theory of integrable systems that we consider in Chapter 10. (For references on this subject, see the bibliographical comments for Chapter 10.) Group theoretic aspects of this theory have been developed in Lu and Weinstein [1990].

For the relationship between Toda flows and gradient flows, and its consequences, see Moser [1975]; Deift *et al.* [1983], and other references in Bloch *et al.* [1992].

Part II
Two-dimensional integrable systems

Chapter 9: Zero-curvature equations

I. Zero-curvature equations.

Let $\mathbf{g}$ be the Lie algebra of a complex Lie group G. (For example, $G = GL_n\mathbf{C}$.) Let $A, B : \mathbf{C} \to \mathbf{g}$. An equation of the form

$$(*) \qquad\qquad A_{\bar{z}} - B_z = [A, B]$$

is called a *zero-curvature equation*.

This should be compared with the Lax equation $(*)$ of Chapter 5, i.e., $\dot{x} = [x, y]$. In fact, it is a generalization of that equation, because we obtain the Lax equation $A_{\bar{z}} = [A, B]$ if $B_z = 0$. (We use complex coordinates $z, \bar{z}$ because of the specific examples which we will introduce later. Some comments on real and complex coordinates appear at the end of this chapter.)

The name "zero-curvature" can be explained as follows. Let $\alpha = Adz + Bd\bar{z}$; this is a 1-form on $\mathbf{C}$ "with values in $\mathbf{g}$". Equation $(*)$ can be written

$$(*) \qquad\qquad d\alpha + \tfrac{1}{2}[\alpha \wedge \alpha] = 0.$$

This is called the Maurer-Cartan equation. If we interpret α as a connection in the (trivial) G-bundle $\mathbf{C} \times G$, then the Maurer-Cartan equation says that the curvature of this connection is zero. This explanation does not play any role in the rest of this book, except in the proof of the next proposition, so the reader who is unfamiliar with the theory of connections should not be unduly disadvantaged.

Proposition. *Let G be a complex Lie group. Let $\alpha = Adz + Bd\bar{z}$, where $A, B : \mathbf{C} \to \mathbf{g}$. The following statements are equivalent:*

(1) α satisfies the equation $d\alpha + \tfrac{1}{2}[\alpha \wedge \alpha] = 0$, i.e., $A_{\bar{z}} - B_z = [A, B]$

(2) There exists a map $F : \mathbf{C} \to G$ such that $\alpha = F^{-1}dF$.

Sketch of the proof. Given (2), we have $A = F^{-1}F_z$, $B = F^{-1}F_{\bar{z}}$. By direct calculation we obtain $A_{\bar{z}} - B_z = [A, B]$. Conversely, if (1) is true, then the horizontal distribution (with respect to the connection α) is integrable. Let M be an integral manifold. Then it can be shown that M is – locally – the graph of a map F (see Chapter 5, section 2, of Sternberg [1964]). Since $\mathbf{C}$ is simply connected, this is true globally. ∎

The function F here is not unique: If $g \in G$, then $(gF)^{-1}d(gF) = F^{-1}dF = \alpha$. However, it is easy to show that F is unique up to left translation by an element of G.

As in the proof of the proposition, statement (2) can be written explicitly in terms of A and B as

$$(**) \qquad\qquad F^{-1}F_z = A, \quad F^{-1}F_{\bar{z}} = B.$$

This should be compared with equation $(**)$ of Chapter 5, i.e., $\dot{u}u^{-1} = -y$. From the point of view of the function F, equation $(**)$ is simpler than equation $(*)$, because it is first order.

The above discussion can also be carried out in the case of a real Lie group. In this case, $\alpha = Adz + Bd\bar{z}$ is a 1-form with values in $\mathbf{g} \otimes \mathbf{C}$. It takes values in the subspace $\mathbf{g} = \{X \in \mathbf{g} \otimes \mathbf{C} \mid c(X) = X\}$ if and only if $B = c(A)$, where the map $c : \mathbf{g} \otimes \mathbf{C} \to \mathbf{g} \otimes \mathbf{C}$ given by $X \otimes z \mapsto X \otimes \bar{z}$ is "conjugation with respect to the real form $\mathbf{g}$". (For example, if $G = U_n$, then $c(X) = -X^*$.) In this case we have:

Proposition. *Let G be a real Lie group. Let $\alpha = Adz + Bd\bar{z}$, where $A, B : \mathbf{C} \to \mathbf{g} \otimes \mathbf{C}$ and $B = c(A)$. The following statements are equivalent:*

(1) α satisfies the equation $d\alpha + \frac{1}{2}[\alpha \wedge \alpha] = 0$, i.e., $A_{\bar{z}} - B_z = [A, B]$

(2) There exists a function $F : \mathbf{C} \to G$ such that $\alpha = F^{-1}dF$. ∎

In Chapter 5, we studied some examples of the equation $\dot{u}u^{-1} = -y$. For certain y, we found the solution explicitly, in terms of an exponential function. It is much more difficult to find all solutions to the above equation $(**)$. However, we might expect that *some* solutions can be obtained in terms of exponentials (and we shall see later that this is so).

II. Example: The two-dimensional Toda lattice.

The *two-dimensional Toda lattice* (or the 2DTL) is the following system of partial differential equations:

$$2(w_i)_{z\bar{z}} = e^{2(w_{i+1} - w_i)} - e^{2(w_i - w_{i-1})}.$$

Here we have $w_i : \mathbf{C} \to \mathbf{R}$ for $i = 0, \dots, n$, and we put

$$w_{-1} = w_n, \qquad w_{n+1} = w_0.$$

We also assume that $\sum_{i=0}^{n} w_i = 0$. (Strictly speaking, this is the *elliptic, periodic* 2DTL.)

This is an obvious generalization of the Toda lattice. (We shall refer to the latter from now on as the one-dimensional Toda lattice, or the 1DTL.) However, it is not an artificial generalization, as it includes some geometrically important examples. For example, when $n = 1$, we have $2(w_1)_{z\bar{z}} = e^{-4w_1} - e^{4w_1}$. Putting $w = 2w_1$, we obtain

$$w_{z\bar{z}} = -2\sinh 2w,$$

which is the "sinh-Gordon equation". This equation is important because its solutions are related to surfaces in $\mathbf{R}^3$ of constant mean curvature. A geometrical interpretation of the general 2DTL is given in Chapter 21.

From our experience with the 1DTL, we now ask the question: Can we express the 2DTL as a matrix equation?

Let $W_{i,j} = e^{w_i - w_j}$. We introduce two matrices A, B as follows:

$$A = \begin{pmatrix} (w_0)_z & 0 & 0 & \cdots & 0 & W_{0,n} \\ W_{1,0} & (w_1)_z & 0 & \cdots & 0 & 0 \\ 0 & W_{2,1} & (w_2)_z & \cdots & 0 & 0 \\ \vdots & \vdots & \vdots & & \vdots & \vdots \\ 0 & 0 & 0 & \cdots & (w_{n-1})_z & 0 \\ 0 & 0 & 0 & \cdots & W_{n,n-1} & (w_n)_z \end{pmatrix}, \quad B = -A^*.$$

A straightforward calculation gives:

Proposition. *The 2DTL is equivalent to the equation $A_{\bar{z}} - B_z = [A, B]$, where A, B are as defined above.* ∎

Furthermore, from section I, we conclude that the 2DTL is equivalent to the system

$$F^{-1}F_z = A, \quad F^{-1}F_{\bar{z}} = B.$$

In other words: If we can find a function $F : \mathbf{C} \to SU_{n+1}$ such that $F^{-1}F_z, F^{-1}F_{\bar{z}}$ have the forms A, B (respectively), then we obtain a solution to the 2DTL. Conversely, any solution of the 2DTL gives rise to such a function F.

III. Example: Harmonic maps from surfaces to Lie groups.

Let G be a compact (real) Lie group. The following discussion is valid for any such G, but to simplify the exposition we shall assume as usual that G is a matrix group.

The equation for a *harmonic map* $\phi : \mathbf{C} \to G$ is the following partial differential equation:

$$(\phi^{-1}\phi_{\bar{z}})_z + (\phi^{-1}\phi_z)_{\bar{z}} = 0.$$

It is well known that this equation is the Euler-Lagrange equation for the functional $\phi \mapsto \int_D ||d\phi||^2$ (where D is a compact domain in $\mathbf{C}$). In other words, $\phi|_D$ satisfies the harmonic map equation if and only if it is a critical point of this functional. Therefore, harmonic maps are a natural generalization of geodesics.

Various important objects in differential geometry are examples of harmonic maps. Some of the most relevant ones for us are minimal surfaces in symmetric spaces, and Gauss maps of surfaces in $\mathbf{R}^3$ of constant mean curvature. We refer the reader to the bibliographical comments at the end of the chapter for further background information on harmonic maps.

If we write $A = \phi^{-1}\phi_z$, $B = \phi^{-1}\phi_{\bar{z}}$, then the harmonic map equation is $A_{\bar{z}} + B_z = 0$. Here we have $A, B : \mathbf{C} \to \mathbf{g} \otimes \mathbf{C}$ with $B = c(A)$. It turns out that the harmonic map equation can be expressed entirely in terms of A and B, without reference to ϕ:

Proposition. *The harmonic map equation is equivalent to the system*

$$A_{\bar{z}} - B_z = [A, B]$$
$$A_{\bar{z}} + B_z = 0$$

where $A, B : \mathbf{C} \to \mathbf{g} \otimes \mathbf{C}$ and $B = c(A)$.

Proof. Let $\phi : \mathbf{C} \to G$ be a harmonic map (i.e., a solution of the harmonic map equation). Let $\alpha = Adz + Bd\bar{z}$, where $A = \phi^{-1}\phi_z$, $B = \phi^{-1}\phi_{\bar{z}}$. Then the first equation follows from section I, and the second equation follows from the harmonic map equation. Conversely, let A, B be solutions of the above system. From the first equation, and section I, there exists $\phi : \mathbf{C} \to G$ (which is unique up to left translation by an element of G) such that $A = \phi^{-1}\phi_z$, $B = \phi^{-1}\phi_{\bar{z}}$. From the second equation, this ϕ is harmonic. ∎

This situation is slightly different from the situation of section II, as we have a zero-curvature equation together with an *extra* equation. However, we can re-write the system in two interesting ways:

(i) The harmonic map equation is equivalent to the system

$$A_{\bar{z}} = [A, \tfrac{1}{2}B]$$
$$B_z = [B, \tfrac{1}{2}A].$$

(Proof: Elementary algebra.) This system consists of two Lax equations.

(ii) The harmonic map equation is equivalent to the equation

$$(A_\lambda)_{\bar{z}} - (B_\lambda)_z = [A_\lambda, B_\lambda] \text{ for all } \lambda \in \mathbf{C} \text{ with } |\lambda| = 1$$

where $A_\lambda = \tfrac{1}{2}(1 - \lambda^{-1})A$, $B_\lambda = \tfrac{1}{2}(1 - \lambda)B$, and $B = c(A)$. (Proof: Elementary algebra, again!)

This equation is a "zero-curvature equation with parameter". By section I, this equation is equivalent to the following system:

$$(**) \qquad F_\lambda^{-1}(F_\lambda)_z = \tfrac{1}{2}(1 - \tfrac{1}{\lambda})A, \quad F_\lambda^{-1}(F_\lambda)_{\bar{z}} = \tfrac{1}{2}(1 - \lambda)B.$$

Observe that it is easy to reconstruct a harmonic map ϕ from F_λ. Namely, $\phi(z) = F_{-1}(z)$. On the other hand, it is not always easy to construct F_λ from ϕ. The map F_λ contains "hidden information" about ϕ.

Note on real and complex derivatives: We are primarily interested in maps $\mathbf{R}^2 \to G$, where G is a real Lie group. However, it is very convenient to identify $\mathbf{R}^2$ with $\mathbf{C}$, and to introduce the complexification G^c of G, because holomorphic maps play a fundamental role in some of our examples. To prevent any possible confusion when switching between the real and complex worlds, we state here some basic definitions and conventions.

First, the identification between $\mathbf{C}$ and $\mathbf{R}^2$ is accomplished by means of the usual formula $z = x + \sqrt{-1}\,y$, and we write $\bar{z} = x - \sqrt{-1}\,y$. The formulae

$$\frac{\partial}{\partial z} = \tfrac{1}{2}\left(\frac{\partial}{\partial x} - \sqrt{-1}\frac{\partial}{\partial y}\right), \quad \frac{\partial}{\partial \bar{z}} = \tfrac{1}{2}\left(\frac{\partial}{\partial x} + \sqrt{-1}\frac{\partial}{\partial y}\right)$$

define complex vector fields on $\mathbf{R}^2$. (Whereas a vector field on a manifold X is a section of the tangent bundle TX, a complex vector field is a section of $TX \otimes \mathbf{C}$.) If V is a real vector space, and $f : \mathbf{C} \to V$ is a function, then we have $f_x, f_y : \mathbf{C} \to V$, but

$$f_z = \tfrac{1}{2}(f_x - \sqrt{-1}\,f_y) : \mathbf{C} \to V \otimes \mathbf{C}$$
$$f_{\bar{z}} = \tfrac{1}{2}(f_x + \sqrt{-1}\,f_y) : \mathbf{C} \to V \otimes \mathbf{C}.$$

For a function $g : \mathbf{C} \to V \otimes \mathbf{C}$, similar definitions of $g_x, g_y, g_z, g_{\bar{z}}$ apply. The Cauchy-Riemann equation, $g_{\bar{z}} = 0$, is the condition for g to be holomorphic.

The complex vector space $V \otimes \mathbf{C}$ has an involution c, namely "conjugation with respect to the real form V". A typical element of $V \otimes \mathbf{C}$ may be written in the form $\sum v_i \otimes z_i$, and by definition $c(\sum v_i \otimes z_i) = \sum v_i \otimes \bar{z}_i$. We have $c(f_z) = f_{\bar{z}}$.

Now suppose that $F : \mathbf{C} \to G$, where G is a real Lie group. We obtain $F^{-1}F_x, F^{-1}F_y : \mathbf{C} \to \mathbf{g}$, and also

$$F^{-1}F_z = \tfrac{1}{2}(F^{-1}F_x - \sqrt{-1}\,F^{-1}F_y) : \mathbf{C} \to \mathbf{g} \otimes \mathbf{C}$$
$$F^{-1}F_{\bar{z}} = \tfrac{1}{2}(F^{-1}F_x + \sqrt{-1}\,F^{-1}F_y) : \mathbf{C} \to \mathbf{g} \otimes \mathbf{C}.$$

If $c : \mathbf{g} \otimes \mathbf{C} \to \mathbf{g} \otimes \mathbf{C}$ is conjugation with respect to the real form $\mathbf{g}$, we have $c(F^{-1}F_z) = F^{-1}F_{\bar{z}}$.

Warning: When we identify a function $f : \mathbf{R}^2 \to V$ with a function $f : \mathbf{C} \to V$, we usually write $f(z)$ instead of $f(z, \bar{z})$. The notation $f(z)$ does not necessarily imply that f is a holomorphic function, however.

Bibliographical comments.

See the comments at the end of Chapter 10.

Chapter 10: Some solutions of zero-curvature equations

It is usually difficult to find the general solution of a zero-curvature equation

$$(*) \qquad\qquad A_{\bar{z}} - B_z = [A, B].$$

In this chapter we describe a well known method of finding *some* non-trivial solutions. The method has two steps:

Step 1: Find a "trivial" solution (e.g., by guessing!).

Step 2: From the trivial solution, construct non-trivial solutions, by using "dressing transformations".

I. Trivial solutions.

As seen in Chapter 9, equation $(*)$ is equivalent to the following system of equations:

$$(**) \qquad\qquad F^{-1}F_z = A, \quad F^{-1}F_{\bar{z}} = B.$$

From the form of these equations, and from our previous experience with Lax equations, we might expect to find solutions of the form "$F = \exp(\ )$". We shall find some solutions for the two examples of Chapter 9.

Examples:

(1) The 2DTL.

The 2DTL equation

$$2(w_i)_{z\bar{z}} = e^{2(w_{i+1} - w_i)} - e^{2(w_i - w_{i-1})}$$

obviously admits the constant solution

$$w_0(z) = \cdots = w_n(z) = 1.$$

The corresponding solution of $(*)$ is given by

$$A(z) = A_0, \quad B(z) = B_0$$

where

$$A_0 = \begin{pmatrix} 0 & 0 & 0 & 0 & 1 \\ 1 & 0 & 0 & 0 & 0 \\ 0 & 1 & 0 & 0 & 0 \\ 0 & 0 & 1 & 0 & 0 \\ 0 & 0 & 0 & 1 & 0 \end{pmatrix}, \quad B_0 = -A_0^* = \begin{pmatrix} 0 & -1 & 0 & 0 & 0 \\ 0 & 0 & -1 & 0 & 0 \\ 0 & 0 & 0 & -1 & 0 \\ 0 & 0 & 0 & 0 & -1 \\ -1 & 0 & 0 & 0 & 0 \end{pmatrix}.$$

A corresponding solution of (∗∗) is given by

$$F(z) = \exp(zA_0 + \bar{z}B_0).$$

Here we use the fact that $[A_0, B_0] = 0$.

(2) Harmonic maps $\mathbf{C} \to G$.

We use the following version of the harmonic map equation (see Chapter 9):

(∗) $(A_\lambda)_{\bar{z}} - (B_\lambda)_z = [A_\lambda, B_\lambda]$ for all $\lambda \in \mathbf{C}$ with $|\lambda| = 1$.

This is equivalent to

(∗∗) $F_\lambda^{-1}(F_\lambda)_z = \frac{1}{2}(1 - \frac{1}{\lambda})A, \quad F_\lambda^{-1}(F_\lambda)_{\bar{z}} = \frac{1}{2}(1 - \lambda)B.$

The original harmonic map equation $(\phi^{-1}\phi_z)_z + (\phi^{-1}\phi_z)_{\bar{z}} = 0$ obviously admits any constant function

$$\phi(z) = g$$

as a solution (where $g \in G$). A corresponding solution of (∗) is simply the zero solution: $A_\lambda = B_\lambda = 0$. This corresponds to a constant solution of (∗∗).

A slightly more interesting solution of (∗∗) is given by

$$F_\lambda(z) = \exp(\tfrac{1}{2}z(1 - \tfrac{1}{\lambda})A_0 + \tfrac{1}{2}\bar{z}(1 - \lambda)B_0)$$

where $A_0, B_0 \in \mathbf{g} \otimes \mathbf{C}$ and $B_0 = c(A_0)$, $[A_0, B_0] = 0$. This is similar to the formula of Example (1) – but in the present case the corresponding harmonic map ϕ is *not* constant (it is given by $\phi(z) = F_{-1}(z) = \exp(zA_0 + \bar{z}B_0)$).

II. Dressing transformations.

The purpose of the "dressing method" is to construct a new solution from an old solution.

We have seen a simple example of such a procedure in Chapter 8, in the case of a Lax equation. Starting with a solution $u : \mathbf{R} \to G_1$ and an element $g \in G_2$, we obtained a new solution $g \cdot u = (gu)_1$. This can be written in a slightly different way, namely $g \cdot u = gu(gu)_2^{-1}$. Since gu is obviously a solution of the same equation as u, we have

new solution $(g \cdot u)$ =   old solution (gu) ×   function $((gu)_2^{-1})$.

We shall examine the harmonic map equation from the same point of view. If F_λ is a solution, we try to obtain a new solution of the form $\tilde{F}_\lambda = F_\lambda G_\lambda$, for some function G_λ. Following the previous example, we try to define $\tilde{F}_\lambda$ by

$$\tilde{F}_\lambda = (g_\lambda F_\lambda)_1 = g_\lambda F_\lambda (g_\lambda F_\lambda)_2^{-1},$$

where g_λ is independent of z. (Thus, we take $G_\lambda = (g_\lambda F_\lambda)_2^{-1}$, essentially.) However, there are two difficulties with this procedure. First, we have not explained the meaning of ( )$_1$. Second, it is certainly not obvious that $\tilde{F}_\lambda$ is a solution. We need a new idea!

We shall make the following (non-trivial) assumption:

Assumption. *There exist $H_\lambda, K_\lambda : \mathbf{C} \to G^c$ such that $g_\lambda F_\lambda = H_\lambda K_\lambda$, and*

(i) H_λ, F_λ are meromorphic in λ on $\mathbf{C} \cup \infty$, and holomorphic in λ on a region $R_1 \subseteq \mathbf{C} \cup \infty$,

(ii) K_λ, g_λ are meromorphic in λ on $\mathbf{C} \cup \infty$, and holomorphic in λ on a region $R_2 \subseteq \mathbf{C} \cup \infty$,

where $R_1 \cup R_2 = \mathbf{C} \cup \infty$ and $0, \infty \in R_2$. In addition, $H_1(z) = F_1(z) = e$ for all z.

We write $H_\lambda = (g_\lambda F_\lambda)_1$, $K_\lambda = (g_\lambda F_\lambda)_2$. This kind of factorization is called a *Riemann-Hilbert factorization*.

Proposition. *Let F_λ be a solution of*

$$(**) \qquad F_\lambda^{-1}(F_\lambda)_z = \tfrac{1}{2}(1 - \tfrac{1}{\lambda})A, \quad F_\lambda^{-1}(F_\lambda)_{\bar{z}} = \tfrac{1}{2}(1 - \lambda)B.$$

Then, under the above Assumption, $\tilde{F}_\lambda = (g_\lambda F_\lambda)_1 = g_\lambda F_\lambda (g_\lambda F_\lambda)_2^{-1}$ is also a solution.

Sketch of the proof. We must calculate $\tilde{F}_\lambda^{-1}(\tilde{F}_\lambda)_z$ and $\tilde{F}_\lambda^{-1}(\tilde{F}_\lambda)_{\bar{z}}$, and show that they have the correct form. We shall just calculate the first one; the calculation of the second is similar. From $\tilde{F}_\lambda = g_\lambda F_\lambda J_\lambda^{-1}$ (where $J_\lambda = (g_\lambda F_\lambda)_2$), we have:

$$\begin{aligned} \tilde{F}_\lambda^{-1}(\tilde{F}_\lambda)_z &= J_\lambda F_\lambda^{-1} g_\lambda^{-1}(g_\lambda(F_\lambda)_z J_\lambda^{-1} + g_\lambda F_\lambda(-J_\lambda^{-1}(J_\lambda)_z J_\lambda^{-1})) \\ &= J_\lambda(F_\lambda^{-1}(F_\lambda)_z)J_\lambda^{-1} - (J_\lambda)_z J_\lambda^{-1} \\ &= \tfrac{1}{2}(1 - \tfrac{1}{\lambda})J_\lambda A J_\lambda^{-1} - (J_\lambda)_z J_\lambda^{-1}. \end{aligned}$$

From this formula, we see that $\tilde{F}_\lambda^{-1}(\tilde{F}_\lambda)_z$ is holomorphic in λ on the region R_2, except for a simple pole at $\lambda = 0$. From the formula $\tilde{F}_\lambda = (g_\lambda F_\lambda)_1$, it follows that $\tilde{F}_\lambda^{-1}(\tilde{F}_\lambda)_z$ is holomorphic in λ on the region R_1. Hence,

$\tilde{F}_\lambda^{-1}(\tilde{F}_\lambda)_z$ is meromorphic on the Riemann sphere $\mathbf{C} \cup \infty$, with a simple pole at $\lambda = 0$. It follows that[4]

$$\tilde{F}_\lambda(z)^{-1}\tilde{F}_\lambda(z)_z = C(z) + \tfrac{1}{\lambda}D(z)$$

for some C, D. For $\lambda = 1$ we obtain 0 (by the last part of the Assumption), hence $C = -D$. We therefore have

$$\tilde{F}_\lambda^{-1}(\tilde{F}_\lambda)_z = \tfrac{1}{2}(1 - \tfrac{1}{\lambda})(2C),$$

as required. ∎

Although the Assumption may appear strange, such a factorization is often possible. Moreover, it can often be obtained from a group decomposition $\mathcal{G} = \mathcal{G}_1\mathcal{G}_2$, where $\mathcal{G}, \mathcal{G}_1, \mathcal{G}_2$ are *infinite dimensional* Lie groups (as we explain in Chapter 12). Thus, this situation is analogous to the situation of the previous example.

Finally, we remark that the same method applies to the 2DTL. For this, we need a version of the 2DTL equation which involves the parameter λ. We define A_λ, B_λ (for $\lambda \in \mathbf{C}^*$) by

$$A_\lambda = \begin{pmatrix} (w_0)_z & 0 & 0 & \cdots & 0 & \lambda W_{0,n} \\ \lambda W_{1,0} & (w_1)_z & 0 & \cdots & 0 & 0 \\ 0 & \lambda W_{2,1} & (w_2)_z & \cdots & 0 & 0 \\ \vdots & \vdots & \vdots & & \vdots & \vdots \\ 0 & 0 & 0 & \cdots & (w_{n-1})_z & 0 \\ 0 & 0 & 0 & \cdots & \lambda W_{n,n-1} & (w_n)_z \end{pmatrix}, \quad B_\lambda = -A_\lambda^*.$$

Exercise:

(10.1) Show that the zero-curvature equation

$$A_{\bar{z}} - B_z = [A, B]$$

(for the 2DTL) is equivalent to the zero-curvature equation

$$(A_\lambda)_{\bar{z}} - (B_\lambda)_z = [A_\lambda, B_\lambda] \text{ for all } \lambda \in \mathbf{C}^*$$

(where A_λ, B_λ are defined above).

This (rather mysterious) fact puts the 2DTL into the same form as the harmonic map equation, i.e., a "zero-curvature equation with parameter".

---

[4]Note that $\tilde{F}_\lambda(z)$, $C(z)$, and $D(z)$ are actually functions of z and $\bar{z}$.

Bibliographical comments for Chapters 9-10.

Like Lax equations, zero-curvature equations appear throughout the literature on integrable systems. An introductory survey is given in Semenov-Tian-Shansky [1982].

Our discussion of the periodic 2DTL here (and later in Chapters 13-15) follows McIntosh [1994a; 1994b]; Bolton *et al.* [1995]. References to earlier work may be found in these articles, and in the article of Fordy in Fordy and Wood [1994], and in Wilson [1981].

The concept of harmonic map makes sense for maps $f : M \to N$ whenever M, N are Riemannian manifolds. Such a map is said to be harmonic if it is a critical point of the energy functional $f \mapsto \int_D ||df||^2$, for all compact D. The form of the Euler-Lagrange equation depends on M and N; we obtain the equation of Chapter 9 if we take the standard Riemannian metric on $M = \mathbf{C}$ and a bi-invariant Riemannian metric on $N = G$. There are several reasons for studying this special case. The most relevant one for us is the existence of a zero-curvature form for the harmonic map equation, in this situation. Other reasons are the special relationship with classical differential geometry (e.g., minimal surfaces), and with mathematical physics (the non-linear sigma model or chiral model).

Good introductions to the general theory of harmonic maps can be found in the lecture notes of Eells and Lemaire [1983] and the book by Urakawa [1993]. The breadth of the subject can be appreciated from the survey articles of Eells and Lemaire [1978; 1988]. All these sources contain information on the applications of harmonic maps in differential geometry and physics.

The zero-curvature form of the harmonic map equation first appeared in Pohlmeyer [1976]. The method of dressing transformations (also called the vesture method) is a standard technique of the theory of integrable systems. This method appeared in the context of harmonic maps in Zakharov and Mikhailov [1978]; Zakharov and Shabat [1979]; a general survey of the method appears in the book by Faddeev and Takhtajan [1987].

The article of Uhlenbeck [1989] was the first comprehensive treatment in the mathematical literature of the zero-curvature equation and dressing transformations for harmonic maps. We discuss this theory in Chapters 16-22.

Chapter 11: Loop groups and loop algebras

I. Loop groups.

Let G be a Lie group. Let $S^1 = \{\lambda \in \mathbf{C} \mid |\lambda| = 1\}$.

Definition.

(1) $\Lambda G = \{\gamma : S^1 \to G \mid \gamma \text{ is smooth}\}$

(2) $\Omega G = \{\gamma : S^1 \to G \mid \gamma \text{ is smooth}, \gamma(1) = e\}$.

We refer to $\Lambda G, \Omega G$ as *loop groups*. More precisely, ΛG is the *free* loop group of G, and ΩG is the *based* loop group of G.

It is obvious that $\Lambda G, \Omega G$ are groups. The group ΛG is a semi-direct product $G \ltimes \Omega G$: As a set, $G \ltimes \Omega G$ is equal to $G \times \Omega G$, but the group operation is given by

$$(g_1, \gamma_1)(g_2, \gamma_2) = (g_1 g_2, \gamma_1 g_1 \gamma_2 g_1^{-1}).$$

(The isomorphism $\Lambda G \to G \ltimes \Omega G$ is given by $\gamma \mapsto (\gamma(1), \gamma\gamma(1)^{-1})$.) It can be shown that $\Lambda G, \Omega G$ are (infinite dimensional) Lie groups. The model spaces for the smooth manifold structures are the infinite dimensional vector spaces

$\Lambda \mathbf{g} = \{f : S^1 \to \mathbf{g} \mid f \text{ is smooth}\}$

$\Omega \mathbf{g} = \{f : S^1 \to \mathbf{g} \mid f \text{ is smooth}, f(1) = 0\}$.

Indeed, the map $f \mapsto \exp \circ f$ defines a local chart (at the identity loop $e \in \Lambda G$ or ΩG), where $\exp : \mathbf{g} \to G$ is the exponential map for G. (For further discussion of the definition of the manifold structure, we refer to Milnor [1984]; Pressley and Segal [1986].) As a manifold, ΛG is diffeomorphic to $G \times \Omega G$. The connected components of ΛG or ΩG are indexed by $\pi_1 G$ (if G is connected).

It is very convenient to represent an element of $\Lambda \mathbf{g}$ by its Fourier series. More generally, any element f of $\Lambda \mathbf{g} \otimes \mathbf{C}$ has a Fourier series representation

$$f(\lambda) = \sum_{i \in \mathbf{Z}} A_i \lambda^i.$$

The condition for such an f to belong to $\Lambda \mathbf{g}$ is $A_i = c(A_{-i})$ for all i. We continue to assume that G is a matrix Lie group, so any $\gamma \in \Lambda G$ (or ΛG^c) has a Fourier series representation as well.

II. The role of loop groups in Lie theory.

Let $\mathbf{g}$ be the Lie algebra of G. We call $\Lambda \mathbf{g}, \Omega \mathbf{g}$ *loop algebras*. It is obvious that $\Lambda \mathbf{g}, \Omega \mathbf{g}$ are Lie algebras. Because of this (and the earlier discussion), we consider $\Lambda \mathbf{g}, \Omega \mathbf{g}$ to be the Lie algebras of $\Lambda G, \Omega G$.

Despite this suggestive terminology, infinite dimensional Lie theory is not very well developed (compared with finite dimensional Lie theory). One of the difficulties is that there is no correspondence between Lie groups and Lie algebras, in general. However, loop algebras are related to an important general class of Lie algebras, namely the *affine Lie algebras*, for which corresponding Lie groups are known to exist.

We shall now give a "concrete" description of affine Lie algebras and their corresponding Lie groups. There are two kinds, un-twisted and twisted. We shall describe each kind separately. In the rest of this section, we assume that G is a compact semisimple Lie group, with complexification G^c. (We use the definition of complexification which was given in Chapter 2. Thus, the Lie algebra of G^c is $\mathbf{g} \otimes \mathbf{C}$.)

(1) The *(un-twisted) affine Lie algebra associated with* $\mathbf{g} \otimes \mathbf{C}$ is defined as follows. Consider the vector space

$$\mathbf{C} \oplus \mathbf{C} \oplus (\mathbf{g} \otimes \mathbf{C}) \otimes \mathbf{C}[\lambda, \lambda^{-1}]$$

where $\mathbf{C}[\lambda, \lambda^{-1}]$ denotes the (infinite dimensional) vector space of polynomials in λ, λ^{-1}. Let us introduce the notation $c = (1,0)$, $d = (0,1)$ for the standard basis of $\mathbf{C} \oplus \mathbf{C}$. Then we define a Lie bracket operation by

$$[A\lambda^k, B\lambda^l] = [A, B]\lambda^{k+l} + k\delta_{k+l,0}(A, B)c$$

(where the Lie bracket on the right hand side is the Lie bracket of $\mathbf{g} \otimes \mathbf{C}$, and (,) is the Killing form), and

$$[c, d] = 0 \quad [c, A\lambda^k] = 0, \quad [d, A\lambda^k] = kA\lambda^k.$$

This is the required affine Lie algebra.

What is a corresponding Lie group? To find one, we should take a suitable *completion* of the above Lie algebra, for a start. Another potential difficulty is that the complexification procedure (for Lie groups) does not work very well in infinite dimensions, so we should work with *real* Lie groups. We shall therefore have to be satisfied with a real Lie group whose complexified Lie algebra contains the above Lie algebra as a dense subspace.

The construction of this Lie group has three steps. First, we have the loop group ΛG; this "corresponds" to $(\mathbf{g} \otimes \mathbf{C}) \otimes \mathbf{C}[\lambda, \lambda^{-1}]$. Next, we consider the semi-direct product $S^1 \ltimes \Lambda G$, where the group structure is given by

$$(\lambda_1, \gamma_1(\lambda))(\lambda_2, \gamma_2(\lambda)) = (\lambda_1 \lambda_2, \gamma_1(\lambda_2^{-1}\lambda)\gamma_2(\lambda)).$$

This corresponds to the Lie algebra $\mathbf{C}d \oplus (\mathbf{g} \otimes \mathbf{C}) \otimes \mathbf{C}[\lambda, \lambda^{-1}]$. Finally, there is a Lie group with the structure of an S^1-bundle over $S^1 \ltimes \Lambda G$, which corresponds to $\mathbf{C}c \oplus \mathbf{C}d \oplus (\mathbf{g} \otimes \mathbf{C}) \otimes \mathbf{C}[\lambda, \lambda^{-1}])$. The construction of this group is more complicated, and is given in Pressley and Segal [1986].

(2) Let α be an outer automorphism of $\mathbf{g}$, and hence of $\mathbf{g} \otimes \mathbf{C}$. Let k be the order of α (necessarily 1, 2, or 3). Let ω be a primitive k-th root of unity. Then the *twisted affine Lie algebra associated with* $\mathbf{g} \otimes \mathbf{C}$ *and* α is defined to be $\mathbf{C} \oplus \mathbf{C} \oplus ((\mathbf{g} \otimes \mathbf{C}) \otimes \mathbf{C}[\lambda, \lambda^{-1}])_\alpha$ where

$$((\mathbf{g} \otimes \mathbf{C}) \otimes \mathbf{C}[\lambda, \lambda^{-1}])_\alpha = \{ f \in (\mathbf{g} \otimes \mathbf{C}) \otimes \mathbf{C}[\lambda, \lambda^{-1}] \mid \alpha(\gamma(\lambda)) = f(\omega\lambda) \}.$$

It is possible to construct a corresponding Lie group in this case also. Namely, we replace ΛG by the *twisted loop group*

$$(\Lambda G)_\alpha = \{ \gamma \in \Lambda G \mid \alpha(\gamma(\lambda)) = \gamma(\omega\lambda) \}$$

(where α now means the automorphism of G induced by the original α). Then we perform two extensions by S^1, as in the un-twisted case.

Affine Lie algebras are important for another reason: They can be characterized "abstractly" in terms of generalized Cartan matrices (or generalized Dynkin diagrams). This is explained in Kac [1990]. The characterization can in fact be extended to a larger class of Lie algebras, the *Kac-Moody Lie algebras*. For this we refer again to Kac [1990].

The above definition of twisted loop groups (or algebras) makes sense both for real and complex Lie groups (or algebras), and for arbitrary automorphisms (inner or outer). We shall make use of these definitions from now on without further comment. The reason for using only *outer* automorphisms in the above discussion is the following result (see Kac [1990], Chapter 8):

Proposition. *If α is an inner automorphism of* $\mathbf{g}$*, then the twisted Lie algebra* $(\Lambda\mathbf{g})_\alpha$ *is isomorphic to the un-twisted Lie algebra* $\Lambda\mathbf{g}$. ∎

To illustrate all this, we conclude with a simple example.

Example:

Let $G = SU_2$, and let α be the automorphism of G or $\mathbf{g}$ given by conjugation by $\begin{pmatrix} 1 & \\ & -1 \end{pmatrix}$. This α is an (inner) automorphism of order 2, so we take $\omega = -1$ in the definition of a twisted loop group.

Using the Fourier series representation $\gamma(\lambda) = \sum_{i \in \mathbf{Z}} A_i \lambda^i$ of an element $\gamma \in \Lambda SU_2$, we have

$$\alpha(\gamma(\lambda)) = \gamma(-\lambda) \iff \alpha(A_i) = (-1)^i A_i \text{ for all } i$$

$$\iff A_i = \begin{cases} \begin{pmatrix} * & 0 \\ 0 & * \end{pmatrix} & \text{for } i \text{ even} \\ \begin{pmatrix} 0 & * \\ * & 0 \end{pmatrix} & \text{for } i \text{ odd.} \end{cases}$$

Exactly the same analysis applies to the Fourier series representation of an element of the Lie group $(\Lambda SL_2\mathbf{C})_\alpha$, or the Lie algebras $(\Lambda su_2)_\alpha$, $(\Lambda sl_2\mathbf{C})_\alpha$.

Using this, one can check that the map

$$\Lambda su_2 \to (\Lambda su_2)_\alpha, \quad \gamma(\lambda) \mapsto \begin{pmatrix} 1 & \\ & \lambda \end{pmatrix} \gamma(\lambda^2) \begin{pmatrix} 1 & \\ & \lambda \end{pmatrix}^{-1}$$

is an isomorphism of Lie algebras. This verifies the last proposition, in this case.

III. The difference between ΛG and ΩG.

For the rest of this chapter, we assume that G is a compact Lie group.

There is a surprising, and important, difference between ΛG and ΩG. Roughly speaking, ΛG is analogous to a compact Lie group, and ΩG is analogous to a compact generalized flag manifold (i.e., adjoint orbit of a compact Lie group). (The definitions of these terms are given in section I of Chapter 3 and section II of Chapter 4.)

The Lie algebra $\sqrt{-1}\,\mathbf{R} \ltimes \Lambda\mathbf{g}$ may be considered as the Lie algebra of $S^1 \ltimes \Lambda G$. The formula

$$\mathrm{Ad}(\lambda,\gamma)\,(\sqrt{-1}\,x,f) = \tfrac{d}{dt}\,(\lambda,\gamma)(e^{\sqrt{-1}\,xt}, \exp tf)(\lambda,\gamma)^{-1}|_0 \quad (\lambda = e^{\sqrt{-1}\,t})$$

defines an action of $S^1 \ltimes \Lambda G$ on $\sqrt{-1}\,\mathbf{R} \ltimes \Lambda\mathbf{g}$. This may be considered as the adjoint representation of $S^1 \ltimes \Lambda G$.

Proposition. *The orbit $M_{(\sqrt{-1},0)}$ of $(\sqrt{-1},0) \in \sqrt{-1}\,\mathbf{R} \ltimes \Lambda\mathbf{g}$ may be identified with ΩG.*

Proof. It is easy to verify that the isotropy subgroup at $(\sqrt{-1},0)$ is $S^1 \times G$. Hence $M_{(\sqrt{-1},0)} \cong (S^1 \ltimes \Lambda G)/(S^1 \times G) \cong \Lambda G/G \cong \Omega G$, as required. ∎

More generally, for any $X \in \mathbf{g}$, the orbit of $(\sqrt{-1}, X)$ is

$$M_{(\sqrt{-1},X)} \cong (S^1 \ltimes \Lambda G)/(S^1 \times C(S_X)) \cong \Lambda G/C(S_X)$$

where $C(S_X)$ denotes the centralizer of the torus S_X which is generated by X (as in Chapter 4).

Bibliographical comments.

See the comments at the end of Chapter 12.

Chapter 12: Factorizations and homogeneous spaces

I. Decomposition of loop groups, and homogeneous spaces.

The Gram-Schmidt decomposition

$$GL_n\mathbf{C} = U_n\Delta_n$$

(where Δ_n denotes the subgroup of $GL_n\mathbf{C}$ consisting of upper triangular matrices; see Chapter 3) generalizes to a decomposition of $\Lambda GL_n\mathbf{C}$. The decomposition is

(†)
$$\Lambda GL_n\mathbf{C} = \Omega U_n \, \Lambda_+ GL_n\mathbf{C},$$

where

$$\Lambda_+ GL_n\mathbf{C} = \{\gamma \in \Lambda GL_n\mathbf{C} \mid \gamma \text{ extends holomorphically for } |\lambda| \leq 1\}.$$

(The condition on γ means that $\gamma = \tilde{\gamma}|_{S^1}$, where $\tilde{\gamma} : \{\lambda \in \mathbf{C} \mid |\lambda| \leq 1\} \to GL_n\mathbf{C}$ is continuous for $|\lambda| \leq 1$ and holomorphic for $|\lambda| < 1$. In terms of the Fourier series representation $\gamma(\lambda) = \sum_{i \in \mathbf{Z}} A_i \lambda^i$, the condition simply means that $A_i = 0$ for all $i < 0$.)

Let us define

$$\Lambda_+^\Delta GL_n\mathbf{C} = \{\gamma \in \Lambda_+ GL_n\mathbf{C} \mid \gamma(1) \in \Delta_n\}.$$

Then, using the facts that $\Lambda U_n = \Omega U_n \, U_n$ and $\Lambda_+ GL_n\mathbf{C} = U_n \Lambda_+^\Delta GL_n\mathbf{C}$, we see that (†) is equivalent to the following decomposition:

(††)
$$\Lambda GL_n\mathbf{C} = \Lambda U_n \, \Lambda_+^\Delta GL_n\mathbf{C}.$$

In principle, (†) can be proved by a generalization of the Gram-Schmidt procedure. However, it is better to take a geometrical approach, as in Chapter 3. Namely, we introduce "flag manifolds" $\mathrm{Gr}^{(n)}, \mathrm{Fl}^{(n)}$ such that

$$\mathrm{Gr}^{(n)} \cong \Lambda U_n/U_n \cong \Lambda GL_n\mathbf{C}/\Lambda_+ GL_n\mathbf{C}$$
$$\mathrm{Fl}^{(n)} \cong \Lambda U_n/T_n \cong \Lambda GL_n\mathbf{C}/\Lambda_+^\Delta GL_n\mathbf{C}$$

where T_n is the subgroup of U_n consisting of diagonal matrices. Before we can define these manifolds, we need to introduce some notation.

Let $H^{(n)}$ be the Hilbert space $L^2(S^1, \mathbf{C}^n)$. If $e_1, \ldots, e_n$ denote the standard basis vectors for $\mathbf{C}^n$, then the functions

$$\lambda \mapsto \lambda^i e_j, \quad i \in \mathbf{Z}, \ j = 1, \ldots, n$$

are a basis for $H^{(n)}$. In other words,

$$H^{(n)} = \text{Span}\{\lambda^i e_j \mid i \in \mathbf{Z}, j = 1, \ldots, n\}$$

where "Span$\{X\}$" means "the closed subspace generated by X". Those functions with $i \geq 0$ generate a closed subspace

$$H^{(n)}_+ = \text{Span}\{\lambda^i e_j \mid i \geq 0, j = 1, \ldots, n\}$$

of $H^{(n)}$. Let $\text{Grass}(H^{(n)})$ denote the space of all vector subspaces $W \subseteq H^{(n)}$ such that

a) W is closed,

b) the projection map $W \to H^{(n)}_+$ is Fredholm, and the projection map $W \to (H^{(n)}_+)^\perp$ is Hilbert-Schmidt,

c) the images of the projection maps $W^\perp \to H^{(n)}_+$, $W \to (H^{(n)}_+)^\perp$ are contained in $C^\infty(S^1, \mathbf{C}^n)$.

The meaning of "Fredholm" and "Hilbert-Schmidt" can be found in books on functional analysis; we do not use these concepts explicitly, so we omit any further explanation of them.

The definitions of $\text{Gr}^{(n)}, \text{Fl}^{(n)}$ are as follows.

Definition.

(1) $\text{Gr}^{(n)} = \{W \in \text{Grass}(H^{(n)}) \mid \lambda W \subseteq W\}$.

(2) $\text{Fl}^{(n)} = \{W_n \subseteq W_{n-1} \subseteq \cdots \subseteq W_1 \subseteq W_0 \mid W_0 \in \text{Gr}^{(n)}, \lambda W_0 = W_n, \dim W_i/W_{i+1} = 1\}$.

Note: If $W \in \text{Gr}^{(n)}$, then it is known that $\dim W/\lambda W = n$.

One of the basic results of Pressley and Segal [1986] is:

Theorem 1 (Grassmannian model of ΩU_n).

$$\text{Gr}^{(n)} \cong \Lambda U_n/U_n \cong \Lambda GL_n\mathbf{C}/\Lambda_+ GL_n\mathbf{C}$$

(and hence $\text{Gr}^{(n)} \cong \Omega U_n$). ∎

The method of proof is to show that $\Lambda GL_n\mathbf{C}$ acts on $\text{Gr}^{(n)}$ in a natural way, and that the subgroup ΛU_n acts *transitively*. The formula for the action is simply

$$\gamma.W = \{\gamma f \mid f \in W\}.$$

By considering Fourier series, it is easy to see that the isotropy subgroup at $H_+^{(n)}$ is $\Lambda_+ GL_n\mathbf{C}$ (for the action of $\Lambda GL_n\mathbf{C}$), or U_n (for the action of the subgroup ΛU_n).

This theorem is a generalization of the (elementary) identifications

$$\mathrm{Gr}_k(\mathbf{C}^n) \cong U_n/(U_k \times U_{n-k}) \cong GL_n\mathbf{C}/P$$

where P denotes the isotropy subgroup of $\mathrm{Span}\{e_1,\ldots,e_k\} \in \mathrm{Gr}_k(\mathbf{C}^n)$ under the natural action of $GL_n\mathbf{C}$.

Regarding the space $\mathrm{Fl}^{(n)}$, there is a similar result:

Theorem 2. $\mathrm{Fl}^{(n)} \cong \Lambda U_n/T_n \cong \Lambda GL_n\mathbf{C}/\Lambda_+^\Delta GL_n\mathbf{C}.$ ∎

The proof is similar to the proof of Theorem 1. This time, the isotropy groups are calculated at the following element of $\mathrm{Fl}^{(n)}$:

$$\lambda H_+^{(n)} \subseteq \mathrm{Span}\{e_1\} \oplus \lambda H_+^{(n)} \subseteq \ldots$$

$$\ldots \subseteq \mathrm{Span}\{e_1,\ldots,e_{n-1}\} \oplus \lambda H_+^{(n)} \subseteq \mathbf{C}^n \oplus \lambda H_+^{(n)} = H_+^{(n)}.$$

This theorem is a generalization of the identifications

$$F_{1,2,\ldots,n-1}(\mathbf{C}^n) \cong U_n/T_n \cong GL_n\mathbf{C}/\Delta_n$$

(which are discussed in Chapter 3).

From the end of the previous chapter, we see that $\mathrm{Gr}^{(n)} \cong M_{(\sqrt{-1},0)}$, and also that $\mathrm{Fl}^{(n)} \cong M_{(\sqrt{-1},X)}$ if X has distinct eigenvalues. These are generalizations of the corresponding facts for $\mathrm{Gr}_k(\mathbf{C}^n)$ and $F_{1,2,\ldots,n-1}(\mathbf{C}^n)$, which we describe in section II of Chapter 4.

The decompositions (†), (††) follow immediately from Theorem 1 and Theorem 2. (Note that $\Lambda U_n \cap \Lambda_+ GL_n\mathbf{C} = U_n$ and $\Lambda U_n \cap \Lambda_+^\Delta GL_n\mathbf{C} = T_n$.) We can refine these decompositions, using the Iwasawa decomposition

$$GL_n\mathbf{C} = U_n A_n N_n$$

of Chapter 3. (Recall that A_n denotes the group of diagonal matrices with positive diagonal entries, and that N_n denotes the subgroup of $GL_n\mathbf{C}$ consisting of matrices of the form

$$\begin{pmatrix} 1 & * & * & * & * \\ & 1 & * & * & * \\ & & 1 & * & * \\ & & & 1 & * \\ & & & & 1 \end{pmatrix}.$$

Thus, $\Delta_n = T_n A_n N_n$.) We obtain:

(†††) $$\Lambda GL_n \mathbf{C} = \Lambda U_n\, A_n\, \Lambda_+^N GL_n \mathbf{C}$$

where $\Lambda_+^N GL_n \mathbf{C} = \{\gamma \in \Lambda_+ GL_n \mathbf{C} \mid \gamma(1) \in N_n\}$. The three subgroups $\Lambda U_n, A_n, \Lambda_+^N GL_n \mathbf{C}$ have (pairwise) trivial intersections, and it can be shown that $\Lambda GL_n \mathbf{C}$ is diffeomorphic to $\Lambda U_n \times A_n \times \Lambda_+^N GL_n \mathbf{C}$.

We refer to any of the three decompositions (†), (††), (†††) as "the Iwasawa decomposition of $\Lambda GL_n \mathbf{C}$".

The results of this section can be generalized to arbitrary compact Lie groups G (see Pressley and Segal [1986]).

II. Riemann-Hilbert problems.

We describe an important analytical interpretation of the Iwasawa decomposition $\Lambda GL_n \mathbf{C} = \Omega U_n \Lambda_+ GL_n \mathbf{C}$. It depends on another (partial) decomposition, called the Birkhoff decomposition.

We begin by describing the finite dimensional analogue of this new decomposition, which is the Gauss decomposition of Chapter 8. (In Chapter 8, we use the "real" version, i.e., for $\mathbf{sl}_n \mathbf{R}$, $SL_n \mathbf{R}$.) The Lie algebra decomposition is

$$(M_n \mathbf{C} =)\ \mathbf{gl}_n \mathbf{C} = \mathbf{n}'_n \oplus \delta_n,$$

where

$$\mathbf{n}'_n = \{\text{strictly lower triangular complex } n \times n \text{ matrices}\}$$
$$\delta_n = \{\text{upper triangular complex } n \times n \text{ matrices}\}.$$

The corresponding partial decomposition of Lie groups is

$$GL_n \mathbf{C} \supseteq N'_n \Delta_n$$

where N'_n denotes the subgroup of $GL_n \mathbf{C}$ consisting of matrices of the form

$$\begin{pmatrix} 1 & & & & \\ * & 1 & & & \\ * & * & 1 & & \\ * & * & * & 1 & \\ * & * & * & * & 1 \end{pmatrix}.$$

The space $N'_n \Delta_n$ is an open dense subspace of $GL_n \mathbf{C}$.

Next we consider the generalization to loop algebras and loop groups.

First we have a Lie algebra decomposition

$$\Lambda \mathbf{gl}_n \mathbf{C} = \Lambda_-^0 \mathbf{gl}_n \mathbf{C} \oplus \Lambda_+ \mathbf{gl}_n \mathbf{C}.$$

where

$$\Lambda_+ \mathbf{gl}_n\mathbf{C} = \mathrm{Span}\{A\lambda^i \mid A \in \mathbf{gl}_n\mathbf{C}, i \geq 0\}$$
$$\Lambda_- \mathbf{gl}_n\mathbf{C} = \mathrm{Span}\{A\lambda^i \mid A \in \mathbf{gl}_n\mathbf{C}, i \leq 0\}$$
$$\Lambda_-^0 \mathbf{gl}_n\mathbf{C} = \{f \in \Lambda_- \mathbf{gl}_n\mathbf{C} \mid f(1) = 0\} = \mathrm{Span}\{A(\lambda^i - 1) \mid A \in \mathbf{gl}_n\mathbf{C}, i < 0\}.$$

The above decomposition is quite elementary.

Then, we have a partial decomposition

$$\Lambda GL_n\mathbf{C} \supseteq \Lambda_-^1 GL_n\mathbf{C} \, \Lambda_+ GL_n\mathbf{C}$$

where

$$\Lambda_- GL_n\mathbf{C} = \{\gamma \in \Lambda GL_n\mathbf{C} \mid \gamma \text{ extends holomorphically for } \infty \geq |\lambda| \geq 1\}$$
$$\Lambda_-^1 GL_n\mathbf{C} = \{\gamma \in \Lambda_- GL_n\mathbf{C} \mid \gamma(1) = I\}.$$

This is called the *Birkhoff decomposition* of $\Lambda GL_n\mathbf{C}$. It is easy to see that $\Lambda_-^1 GL_n\mathbf{C} \cap \Lambda_+ GL_n\mathbf{C} = \{I\}$ (because $\Lambda_- GL_n\mathbf{C} \cap \Lambda_+ GL_n\mathbf{C}$ consists of the loops which extend holomorphically to the entire Riemann sphere $\mathbf{C} \cup \infty$, i.e., the constant loops). It is known that $\Lambda_-^1 GL_n\mathbf{C} \, \Lambda_+ GL_n\mathbf{C}$ is an open dense subspace of the identity component of $\Lambda GL_n\mathbf{C}$. (In fact – see Pressley and Segal [1986], Chapter 8 – any $\gamma \in \Lambda GL_n\mathbf{C}$ can be factorized as $\gamma = \gamma_- \gamma_0 \gamma_+$ with $\gamma_- \in \Lambda_-^1 GL_n\mathbf{C}$, $\gamma_+ \in \Lambda_+ GL_n\mathbf{C}$, and

$$\gamma_0(\lambda) = \begin{pmatrix} \lambda^{k_1} & & \\ & \ddots & \\ & & \lambda^{k_n} \end{pmatrix}$$

for some integers $k_1, \ldots, k_n$. The case $k_1 = \cdots = k_n = 0$ gives all loops in $\Lambda_-^1 GL_n\mathbf{C} \, \Lambda_+ GL_n\mathbf{C}$. We discuss this further in Chapter 17.)

This partial decomposition arises naturally if one considers the following kind of problem:

Riemann-Hilbert Problem. *Let Γ be a (not necessarily connected) simple closed contour in the Riemann sphere $\mathbf{C} \cup \infty$. Let F be a (smooth) matrix-valued function on Γ. When can we find matrix-valued functions F_+, F_- such that*

(1) $F = F_-|_\Gamma \, F_+|_\Gamma$

(2) F_+ (F_-) is holomorphic on the interior (exterior) of Γ?

Examples:

(1) Let

$$\Gamma = S^1 = \{\lambda \mid |\lambda| = 1\}$$
$$F : S^1 \to GL_n\mathbf{C}, \text{ smooth.}$$

We define the interior and exterior of Γ by

$$\text{interior} = D_+ = \{\lambda \mid |\lambda| \le 1\}$$
$$\text{exterior} = D_- = \{\lambda \mid 1 \le |\lambda| \le \infty\}.$$

When do there exist

$$F_+ : D_+ \to GL_n\mathbf{C}, \text{ holomorphic}$$
$$F_- : D_- \to GL_n\mathbf{C}, \text{ holomorphic}$$

such that $F = F_-|_{S^1} F_+|_{S^1}$? The previous discussion tells us that such F_+, F_- exist for "generic" F in the identity component of $\Lambda GL_n\mathbf{C}$.

(2) Let

$$\Gamma = C^\epsilon \cup C^{1/\epsilon}$$
$$F = (F^\epsilon, F^{1/\epsilon}) : C^\epsilon \cup C^{1/\epsilon} \to GL_n\mathbf{C}, \text{ smooth}$$

where $C^r = \{\lambda \mid |\lambda| = r\}$ for $r = \epsilon, 1/\epsilon$, and where we assume $0 < \epsilon < 1$. Thus the contour Γ consists of two disjoint "circles of latitude", one in the northern hemisphere and one in the southern hemisphere.

Let

$$\text{interior} = I = \{\lambda \mid |\lambda| \le \epsilon\} \cup \{\lambda \mid 1/\epsilon \le |\lambda| \le \infty\} = I^\epsilon \cup I^{1/\epsilon}$$
$$\text{exterior} = E = \{\lambda \mid \epsilon \le |\lambda| \le 1/\epsilon\}.$$

When do there exist

$$F_I = (F_I^\epsilon, F_I^{1/\epsilon}) : I^\epsilon \cup I^{1/\epsilon} \to GL_n\mathbf{C}, \text{ holomorphic}$$
$$F_E : E \to GL_n\mathbf{C}, \text{ holomorphic}$$

such that $F^\epsilon = F_E F_I^\epsilon$ on C^ϵ and $F^{1/\epsilon} = F_E F_I^{1/\epsilon}$ on $C^{1/\epsilon}$?

Like Example (1), Example (2) also involves a "partial decomposition", namely

$$\Lambda^{\epsilon,1/\epsilon} GL_n\mathbf{C} \supseteq \Lambda_E^1 GL_n\mathbf{C}\, \Lambda_I GL_n\mathbf{C}$$

where

$$\Lambda^{\epsilon,1/\epsilon}GL_n\mathbf{C} = \{\gamma : C^\epsilon \cup C^{1/\epsilon} \to GL_n\mathbf{C} \mid \gamma \text{ smooth}\} \cong \Lambda GL_n\mathbf{C} \times \Lambda GL_n\mathbf{C}$$

and where $\Lambda_E^1 GL_n\mathbf{C}$, $\Lambda_I GL_n\mathbf{C}$ are defined in the obvious way. It can be shown that the right hand side is an open dense subspace of a component of $\Lambda^{\epsilon,1/\epsilon}GL_n\mathbf{C}$ (see Bergvelt and Guest [1991]).

Example (2) is interesting because the above partial decomposition is related to the Iwasawa decomposition, and this leads to a "complete answer" to the corresponding Riemann-Hilbert problem. To see this, we shall consider a special case of Example (2) where we assume equivariance with respect to the involutions

$$\mathbf{C} \cup \infty \to \mathbf{C} \cup \infty, \quad \lambda \mapsto 1/\bar{\lambda}$$
$$GL_n\mathbf{C} \to GL_n\mathbf{C}, \quad A \mapsto A^{-1*} \ (= (A^{-1})^*).$$

In other words, we make the following assumption:

Reality Assumption. $F(1/\bar{\lambda}) = F(\lambda)^{-1*}$.

(For the purposes of later applications, this turns out to be a very natural assumption.) We then have:

Proposition. *In Example (2), assume that*

(1) $F = (F^\epsilon, F^{1/\epsilon})$ satisfies the Reality Assumption, and

(2) F^ϵ extends holomorphically to $\{\lambda \mid \epsilon \le |\lambda| \le 1\}$ (i.e., $F^\epsilon = G|_{C^\epsilon}$ for some holomorphic $G : \{\lambda \mid \epsilon \le |\lambda| \le 1\} \to GL_n\mathbf{C}$).

Then the required F_E, F_I exist. Moreover, F_E is given by $F_E = G_1$, where $G|_{S^1} = G_1 G_2$ is the factorization of G with respect to the Iwasawa decomposition $\Lambda GL_n\mathbf{C} = \Omega U_n \Lambda_+ GL_n\mathbf{C}$.

Proof. The reality condition is equivalent to the condition

$$F^{1/\epsilon}(\lambda) = F^\epsilon(1/\bar{\lambda})^{-1*}.$$

Since $G|_{S^1}$ and G_2 extend holomorphically to $\{\lambda \mid \epsilon \le |\lambda| \le 1\}$, the same is true of G_1. Since $G_1(\lambda)^* = G_1(\lambda)^{-1}$ for $|\lambda| = 1$, we can further extend G holomorphically to $\{\lambda \mid \epsilon \le |\lambda| \le 1/\epsilon\}$, by defining $G_1(\lambda) = G_1(1/\bar{\lambda})^{-1*}$ for $1 \le |\lambda| \le 1/\epsilon$.

We have

$$(F^\epsilon(\lambda), F^\epsilon(1/\bar{\lambda})^{-1*}) =$$
$$(G_1(\lambda)|_{C^\epsilon}, G_1(1/\bar{\lambda})^{-1*}|_{C^{1/\epsilon}})(G_2(\lambda)|_{C^\epsilon}, G_2(1/\bar{\lambda})^{-1*}|_{C^{1/\epsilon}}).$$

Hence we can take

$$F_E = G_1, \quad F_I = (G_2(\lambda)|_{C^\epsilon}, G_2(1/\bar{\lambda})^{-1*}|_{C^{1/\epsilon}})$$

and the proof is complete. ∎

Hypothesis (2) was included in order to simplify the proof of the proposition. It turns out that the proposition remains true if this hypothesis is dropped. This is a consequence of the fact that the partial decomposition $\Lambda^{\epsilon,1/\epsilon}GL_n\mathbf{C} \supseteq \Lambda^1_E GL_n\mathbf{C}\,\Lambda_I GL_n\mathbf{C}$ becomes a *decomposition*, under the Reality Assumption:

Proposition (McIntosh [1994a], Proposition 6.2). *We have*

$$\Lambda^{\epsilon,1/\epsilon}_{\mathbf{R}}GL_n\mathbf{C} = \Lambda^1_{E,\mathbf{R}}GL_n\mathbf{C}\,\Lambda_{I,\mathbf{R}}GL_n\mathbf{C},$$

where the suffix **R** *indicates that the Reality Assumption is in force.*

Proof. For $(\gamma^\epsilon, \gamma^{1/\epsilon}) \in \Lambda^{\epsilon,1/\epsilon}_{\mathbf{R}}GL_n\mathbf{C}$, we must find $\gamma_E \in \Lambda^1_{E,\mathbf{R}}GL_n\mathbf{C}$ and $\gamma_I \in \Lambda_{I,\mathbf{R}}GL_n\mathbf{C}$ such that $\gamma = \gamma_E\gamma_I$. At the beginning of this section, we mentioned the factorization $\gamma = \gamma_-\gamma_0\gamma_+$ of a loop $\gamma \in \Lambda GL_n\mathbf{C}$. This is easily modified to give a factorization $\gamma^\epsilon = \gamma^\epsilon_-\gamma^\epsilon_0\gamma^\epsilon_+$, where $\gamma^\epsilon_-, \gamma^\epsilon_+$ are defined (respectively) for $|\lambda| \geq \epsilon$, $|\lambda| \leq \epsilon$, and γ^ϵ_0 is defined for all $\lambda \in \mathbf{C}^*$. In addition, we may assume that $\gamma^\epsilon_-(1)\gamma^\epsilon_0(1) = I$.

We seek α, β such that

$$(\gamma^\epsilon, \gamma^{1/\epsilon}) = (\gamma^\epsilon_-\gamma^\epsilon_0\alpha)(\alpha^{-1}\gamma^\epsilon_+, \beta)$$

where α, β are defined for $|\lambda| \leq 1/\epsilon$, $|\lambda| \geq 1/\epsilon$, and $\alpha(1) = I$. This equation is equivalent to $\gamma^{1/\epsilon} = \gamma^\epsilon_-\gamma^\epsilon_0\alpha\beta$, or $(\gamma^\epsilon_0)^{-1}(\gamma^\epsilon_-)^{-1}\gamma^{1/\epsilon} = \alpha\beta$, on the circle $|\lambda| = 1/\epsilon$. When we impose the Reality Assumption $\gamma^{1/\epsilon}(1/\bar{\lambda}) = \gamma^\epsilon(\lambda)^{-1*}$, we claim that we can find such α and β. To do this, we re-write the above equation (on $|\lambda| = 1/\epsilon$) as

$$\gamma^\epsilon_0(\lambda)^{-1}\gamma^\epsilon_-(\lambda)^{-1}\gamma^\epsilon_-(1/\bar{\lambda})^{-1*}\gamma^\epsilon_0(1/\bar{\lambda})^{-1*} = \alpha(\lambda)\beta(\lambda)\gamma^\epsilon_+(1/\bar{\lambda})^*.$$

The left hand side is $\delta(1/\bar{\lambda})^*\delta(\lambda)$, where $\delta(\lambda) = \gamma^\epsilon_-(1/\bar{\lambda})^{-1*}\gamma^\epsilon_0(1/\bar{\lambda})^{-1*}$. Observe that δ is defined for $\epsilon \leq |\lambda| \leq 1/\epsilon$. Let $\delta = \delta_1\delta_2$ be the factorization of δ with respect to the Iwasawa decomposition $\Lambda GL_n\mathbf{C} = \Lambda U_n\,\Lambda^1_-GL_n\mathbf{C}$. We have

$$\delta(1/\bar{\lambda})^*\delta(\lambda) = \delta_2(1/\bar{\lambda})^*\delta_1(1/\bar{\lambda})^*\delta_1(\lambda)\delta_2(\lambda)$$
$$= \delta_2(1/\bar{\lambda})^*\delta_2(\lambda).$$

Here, δ_2 is defined for $|\lambda| \geq 1$, and in particular for $|\lambda| \geq 1/\epsilon$. A priori, $\delta_2(1/\bar{\lambda})^*$ is defined for $|\lambda| \leq 1$, but it must extend to the region $1 \leq |\lambda| \leq 1/\epsilon$, because of the above equation (and the fact that δ_2, δ both extend to $1 \leq |\lambda| \leq 1/\epsilon$). We may therefore take $\alpha(\lambda) = \delta_2(1/\bar{\lambda})^*$, $\beta(\lambda)\gamma^\epsilon_+(1/\bar{\lambda})^* = \delta_2(\lambda)$. It is easy to verify that the resulting γ_E and γ_I satisfy the Reality Assumption. ∎

Bibliographical comments for Chapters 11-12.

An introduction to infinite dimensional Lie groups is given in Milnor [1984]. The standard reference for loop groups is the book by Pressley and Segal [1986]; the article of Freed [1988] is also a useful reference. The standard reference for affine Lie algebras (and, more generally, Kac-Moody Lie algebras) is the book by Kac [1990].

The idea of using infinite dimensional Grassmannians in the theory of integrable systems is due to Sato (see Sato [1981]). For the loop group theoretic point of view, see Goodman and Wallach [1984a]; Wilson [1985]; Segal and Wilson [1985]. In the following chapters we are concerned with the application of this theory to the 2DTL and the harmonic map equation.

Chapter 13: The two-dimensional Toda lattice

I. The 2DTL, again.

Recall from Chapter 9 that the (elliptic, periodic) two-dimensional Toda lattice (or the 2DTL) is the system of equations

$$2(w_i)_{z\bar{z}} = e^{2(w_{i+1}-w_i)} - e^{2(w_i-w_{i-1})}, \quad i = 0, \ldots, n$$

where $w_i : \mathbf{C} \to \mathbf{R}$, $w_{-1} = w_n$, $w_{n+1} = w_0$, and $\sum_{i=0}^{n} w_i = 0$.

From Chapter 9 we know also that the 2DTL is equivalent to the zero-curvature equation

$$A_{\bar{z}} - B_z = [A, B]$$

where

$$A = \begin{pmatrix} (w_0)_z & 0 & 0 & \cdots & 0 & W_{0,n} \\ W_{1,0} & (w_1)_z & 0 & \cdots & 0 & 0 \\ 0 & W_{2,1} & (w_2)_z & \cdots & 0 & 0 \\ \vdots & \vdots & \vdots & & \vdots & \vdots \\ 0 & 0 & 0 & \cdots & (w_{n-1})_z & 0 \\ 0 & 0 & 0 & \cdots & W_{n,n-1} & (w_n)_z \end{pmatrix} = -B^*$$

and $W_{i,i-1} = e^{w_i - w_{i-1}}$.

Moreover, because of the geometrical interpretation of zero-curvature equations, we know that the 2DTL is equivalent to the system

$$F^{-1}F_z = A$$
$$F^{-1}F_{\bar{z}} = B$$

where $F : \mathbf{C} \to SU_{n+1}$. The second equation here is equivalent to the first equation: It may be obtained from the first equation by applying the involution $X \mapsto -X^*$ of $\mathrm{sl}_{n+1}\mathbf{C}$. It is for this reason that F is SU_{n+1}-valued, rather than $SL_{n+1}\mathbf{C}$-valued.

From Chapter 10 we know that the zero-curvature equation

$$A_{\bar{z}} - B_z = [A, B]$$

is equivalent to the zero-curvature equation

$$(A_\lambda)_{\bar{z}} - (B_\lambda)_z = [A_\lambda, B_\lambda] \quad \text{for all } \lambda \in \mathbf{C} \text{ with } |\lambda| = 1$$

where A_λ, B_λ are defined as follows:

$$A_\lambda = \begin{pmatrix} (w_0)_z & 0 & 0 & \cdots & 0 & \lambda W_{0,n} \\ \lambda W_{1,0} & (w_1)_z & 0 & \cdots & 0 & 0 \\ 0 & \lambda W_{2,1} & (w_2)_z & \cdots & 0 & 0 \\ \vdots & \vdots & \vdots & & \vdots & \vdots \\ 0 & 0 & 0 & \cdots & (w_{n-1})_z & 0 \\ 0 & 0 & 0 & \cdots & \lambda W_{n,n-1} & (w_n)_z \end{pmatrix} = -B_\lambda^*.$$

Thus, the "parameter" λ may be inserted, without changing the problem.

Surprisingly, this "unnecessary parameter" is the key to finding solutions of the 2DTL. This is because the 2DTL admits a very large symmetry group – namely a loop group. In the original formulation of the 2DTL (above), this symmetry group is not visible. It is revealed only when we include the parameter λ.

To explain this, we must

(1) identify the appropriate loop group, and

(2) show that it acts on the space of solutions of the 2DTL.

These are the aims of this chapter.

II. The loop group formulation of the 2DTL.

From the geometrical interpretation of zero-curvature equations, we know that the 2DTL is equivalent to the system

$$F_\lambda^{-1}(F_\lambda)_z = A_\lambda$$
$$F_\lambda^{-1}(F_\lambda)_{\bar{z}} = B_\lambda$$

where $F_\lambda : \mathbf{C} \to SU_{n+1}$. By "equivalent" we mean: If we can find F_λ such that $F_\lambda^{-1}(F_\lambda)_z, F_\lambda^{-1}(F_\lambda)_{\bar{z}}$ have the forms A_λ, B_λ (respectively), then we obtain a solution to the 2DTL. Conversely, any solution of the 2DTL gives rise to such a function F_λ. It is clear how the loop group ΛSU_{n+1} enters the picture, because F_λ corresponds to a map

$$F : \mathbf{C} \to \Lambda SU_{n+1}, \quad F(z)(\lambda) = F_\lambda(z).$$

Notation: From now on, F will always denote the ΛSU_{n+1}-valued map corresponding to F_λ, so there should be no possibility of confusion with the earlier meaning of F. Similarly, we shall denote by A, B the $\Lambda \mathfrak{sl}_{n+1}\mathbf{C}$-valued maps corresponding to A_λ, B_λ. To simplify the notation further, we shall write $F(z, \lambda)$ instead of $F(z)(\lambda)$, and similarly for A, B.

In fact, it is better to consider a certain subgroup of ΛSU_{n+1}. Recall that the twisted loop group $(\Lambda SU_{n+1})_\sigma$ is defined as follows:

$$(\Lambda SU_{n+1})_\sigma = \{\gamma \in \Lambda SU_{n+1} \mid \gamma(\omega\lambda) = \sigma(\gamma(\lambda))\}$$

where $\omega = e^{2\pi\sqrt{-1}/k}$ and σ is an automorphism of order k. We choose σ to be the (inner) automorphism

$$\sigma \quad = \quad \text{conjugation by} \quad \begin{pmatrix} 1 & & & \\ & \omega & & \\ & & \ddots & \\ & & & \omega^n \end{pmatrix} \quad (= \hat{\omega}, \text{ say})$$

where $\omega = e^{2\pi\sqrt{-1}/(n+1)}$.

The Lie algebra $(\Lambda su_{n+1})_\sigma$ of $(\Lambda SU_{n+1})_\sigma$ has a simple description. The eigenvalues of σ on $\mathbf{g} = \mathbf{sl}_{n+1}\mathbf{C}$ are $1, \omega, \ldots, \omega^n$. The ω^r-eigenspace, which we denote by $\mathbf{g}_r$, is

$$\mathbf{g}_r = \{(a_{ij})_{0 \le i,j \le n} \in \mathbf{sl}_{n+1}\mathbf{C} \mid a_{ij} = 0 \text{ if } i \ne j + r \bmod n + 1\}.$$

In other words, the eigenspaces

$$\mathbf{g}_0, \ \mathbf{g}_1, \ \cdots \ , \mathbf{g}_n$$

consist (respectively) of matrices like this:

$$\begin{pmatrix} * & 0 & 0 & 0 & 0 \\ 0 & * & 0 & 0 & 0 \\ 0 & 0 & * & 0 & 0 \\ 0 & 0 & 0 & * & 0 \\ 0 & 0 & 0 & 0 & * \end{pmatrix}, \begin{pmatrix} 0 & 0 & 0 & 0 & * \\ * & 0 & 0 & 0 & 0 \\ 0 & * & 0 & 0 & 0 \\ 0 & 0 & * & 0 & 0 \\ 0 & 0 & 0 & * & 0 \end{pmatrix}, \cdots, \begin{pmatrix} 0 & * & 0 & 0 & 0 \\ 0 & 0 & * & 0 & 0 \\ 0 & 0 & 0 & * & 0 \\ 0 & 0 & 0 & 0 & * \\ * & 0 & 0 & 0 & 0 \end{pmatrix}.$$

We can extend the notation $\mathbf{g}_r$ in a convenient way, by defining $\mathbf{g}_{r+j(n+1)} = \mathbf{g}_r$ for any $j \in \mathbf{Z}$.

It follows that the twisted Lie algebra $(\Lambda sl_{n+1}\mathbf{C})_\sigma$ is given by the subalgebra

$$\text{Span}\{A_i \lambda^i \mid A_i \in \mathbf{g}_i, i \in \mathbf{Z}\}$$

of $\Lambda sl_{n+1}\mathbf{C} = \text{Span}\{A\lambda^i \mid A \in \mathbf{g}, i \in \mathbf{Z}\}$. (For the case $n = 1$, see section II of Chapter 11.)

The relevance of the twisted loop group to our problem is that the maps

$$A, B : \mathbf{C} \to \Lambda sl_{n+1}\mathbf{C}$$

take values in the twisted subalgebra $(\Lambda sl_{n+1}\mathbf{C})_\sigma$ (see the formulae for A_λ, B_λ at the beginning of this chapter). Hence the map F takes values in the twisted subgroup $(\Lambda SU_{n+1})_\sigma$.

We are led to consider the following conditions on a map $F : \mathbf{C} \to (\Lambda SU_{n+1})_\sigma$:

(Λ_σ)
$$F^{-1}F_z = \text{linear in } \lambda$$
$$F^{-1}F_{\bar{z}} = \text{linear in } \tfrac{1}{\lambda}.$$

Any solution of the 2DTL gives a solution of (Λ_σ). Conversely, if F is a solution of (Λ_σ), then we have

$$F^{-1}F_z = \begin{pmatrix} a_0 & & & \\ & a_1 & & \\ & & \cdots & \\ & & & \cdots \\ & & & & a_n \end{pmatrix} + \lambda \begin{pmatrix} 0 & & & & b_0 \\ b_1 & 0 & & & \\ & \cdots & \cdots & & \\ & & \cdots & \cdots & \\ & & & b_n & 0 \end{pmatrix}$$

(and $F^{-1}F_{\bar{z}} = -(F^{-1}F_z)^*$), where $a_0, \ldots, a_n, b_0, \ldots, b_n \in \mathbf{C}$. In addition, we have various conditions on a_i and b_i from the identity

$$(F^{-1}F_z)_{\bar{z}} - (F^{-1}F_{\bar{z}})_z = [F^{-1}F_z, F^{-1}F_{\bar{z}}].$$

These conditions are given by writing down the coefficients of $\lambda, 1, \lambda^{-1}$:

(λ) $\qquad\qquad\qquad (b_i)_{\bar{z}} = b_i(\bar{a}_i - \bar{a}_{i-1})$

(1) $\qquad\qquad\qquad (a_i)_{\bar{z}} + (\bar{a}_i)_z = |b_{i+1}|^2 - |b_i|^2$

(λ^{-1}) $\qquad\qquad\qquad (\bar{b}_i)_z = \bar{b}_i(a_i - a_{i-1}).$

Let us now impose the additional conditions

(T) $\qquad\qquad\qquad b_0, \ldots, b_n > 0, \quad b_0 \ldots b_n = 1.$

Equation (λ^{-1}) becomes $(\log b_i)_z = a_i - a_{i-1}$. This may be written in matrix form as $L_z = A - \mathrm{Ad}(E)A$, where L is diagonal with diagonal terms $\log b_0, \ldots, \log b_n$, A is diagonal with diagonal terms $a_0, \ldots, a_n$, and

$$E = \begin{pmatrix} 0 & & & 1 \\ 1 & 0 & & \\ & \ddots & \ddots & \\ & & \ddots & \ddots \\ & & & 1 & 0 \end{pmatrix}.$$

It is easy to verify that the linear transformation $I - \mathrm{Ad}\,E$ on $(\mathbf{sl}_{n+1}\mathbf{C})_\sigma$ is invertible, with inverse $\sum_{i=0}^{n-1}(n-i)\,\mathrm{Ad}\,E^i/(n+1)$. Hence, $a_i = (w_i)_z$ for some real function w_i, and $(\log b_i)_z = (w_i - w_{i-1})_z$. Since b_i and w_i are real, we obtain $b_i = u_i e^{w_i - w_{i-1}}$ for some positive constants u_i. We can arrange that $u_i = 1$ for all i, by replacing F by

$$F \begin{pmatrix} 1 & & & \\ & e_1 & & \\ & & \ddots & \\ & & & \ddots & \\ & & & & e_n \end{pmatrix}$$

where $e_i = (u_1 \ldots u_i)^{-1}$. We end up with

$$a_i = (w_i)_z, \quad b_i = e^{w_i - w_{i-1}}$$

which is precisely the form of the 2DTL.

We have therefore arrived at a loop group theoretic form of the 2DTL, which is independent of the original functions w_i. Symbolically:

$$\text{2DTL} \iff (\Lambda_\sigma) + (T).$$

Note on the parameter λ: It is interesting to compare the zero-curvature equation

$$(Z) \qquad\qquad A_{\bar{z}} - B_z = [A, B]$$

where

$$A = \begin{pmatrix} a_0 & & & \\ & a_1 & & \\ & & \ddots & \\ & & & \ddots \\ & & & & a_n \end{pmatrix} + \begin{pmatrix} 0 & & & b_0 \\ b_1 & 0 & & \\ & \ddots & \ddots & \\ & & \ddots & \ddots \\ & & & b_n & 0 \end{pmatrix} = -B^*$$

with the "zero-curvature equation with parameter"

$$(Z_\lambda) \qquad\qquad (A_\lambda)_{\bar{z}} - (B_\lambda)_z = [A_\lambda, B_\lambda]$$

where

$$A_\lambda = \begin{pmatrix} a_0 & & & \\ & a_1 & & \\ & & \ddots & \\ & & & \ddots \\ & & & & a_n \end{pmatrix} + \lambda \begin{pmatrix} 0 & & & b_0 \\ b_1 & 0 & & \\ & \ddots & \ddots & \\ & & \ddots & \ddots \\ & & & b_n & 0 \end{pmatrix} = -B_\lambda^*.$$

As we have seen, (Z_λ) is equivalent to equations (λ), (1) and (λ^{-1}). Equation (Z) turns out to be equivalent to these three equations as well, when $n > 1$. But for $n = 1$, (Z) is equivalent to:

(1) $\qquad\qquad (a_i)_{\bar{z}} + (\bar{a}_i)_z = |b_{i+1}|^2 - |b_i|^2$

(2) $\qquad\qquad (\bar{b}_i)_z + (b_{i+1})_{\bar{z}} = \bar{b}_i(a_i - a_{i-1}) + b_{i+1}(\bar{a}_{i+1} - \bar{a}_i)$

(equation (2) is obtained by adding the $(i + 1)$-th equation of (λ) to the i-th equation of (λ^{-1})).

In the special case where $a_i = (w_i)_z$ and $b_i = e^{w_i - w_{i-1}}$ – i.e., in the presence of condition (T) – then equations (λ) and (λ^{-1}) become identities. So in this case we see that (Z_λ) is equivalent to (Z) for all n.

III. The symmetry group of the 2DTL.

A method of constructing symmetries has already been suggested in Chapter 10. Let us review that method, using the language of loop groups this time. Let $\gamma \in (\Lambda SL_{n+1}\mathbf{C})_\sigma$ be a loop, and let $F : \mathbf{C} \rightarrow (\Lambda SU_{n+1})_\sigma$ be a solution of

$$(\Lambda_\sigma) \qquad\qquad F^{-1}F_z = \text{linear in } \lambda$$
$$F^{-1}F_{\bar{z}} = \text{linear in } \tfrac{1}{\lambda}.$$

(We ignore condition (T) in this section.) Then we consider a new map

$$\tilde{F} = \gamma \cdot F = (\gamma F)_1.$$

This means: We *assume* that γF can be factorized in the form $\gamma F = (\gamma F)_1 (\gamma F)_2$, where $(\gamma F)_1$ extends holomorphically (in λ) to a region R_1, and $(\gamma F)_2$ extends holomorphically (in λ) to a region R_2, where R_1 and R_2 are as in Chapter 10.

We claim:

(1) $\gamma\delta \cdot F = \gamma \cdot (\delta \cdot F)$ for all $\gamma, \delta \in (\Lambda SL_{n+1}\mathbf{C})_\sigma$ (i.e., we have a group action), and

(2) $\gamma \cdot F$ is a solution of (Λ_σ)

(assuming that all relevant factorizations are possible).

The proof of (1) is elementary algebra:

$$\gamma\delta \cdot F = (\gamma\delta F)_1 = (\gamma(\delta F)_1)_1 = \gamma \cdot (\delta F)_1 = \gamma \cdot (\delta \cdot F)$$

(as in section II of Chapter 8).

To prove (2), we use the argument of the proposition of Chapter 10. Namely, we observe that (a) $\tilde{F}^{-1}\tilde{F}_z$ is holomorphic on R_1 (by definition of $\tilde{F}$), and (b) $\tilde{F}^{-1}\tilde{F}_z$ has the same poles as $F^{-1}F_z$ on R_2 (by substituting $\tilde{F} = \gamma F(\gamma F)_2^{-1}$ and doing a short calculation). It follows that $\tilde{F}^{-1}\tilde{F}_z$ has the same poles as $F^{-1}F_z$ on the entire Riemann sphere $\mathbf{C}\cup\infty$. Now, $F^{-1}F_z$ has a simple pole at $\lambda = \infty$, and no other poles. Therefore, the same is true of $\tilde{F}^{-1}\tilde{F}_z$ – and this means that $\tilde{F}^{-1}\tilde{F}_z$ is linear in λ. A similar argument shows that $\tilde{F}^{-1}\tilde{F}_{\bar{z}}$ is linear in λ^{-1}.

We have now proved (1) and (2), under the assumption concerning the factorization. Let us now examine this assumption, and in particular the nature of the regions R_1 and R_2.

It is tempting to choose $R_1 = D_-$ and $R_2 = D_+$, because the domain of our loops is precisely $D_- \cap D_+ = S^1$. However, it is much better to choose the regions

$$R_1 = E = \{\lambda \mid \epsilon \leq |\lambda| \leq 1/\epsilon\}$$
$$R_2 = I = \{\lambda \mid |\lambda| \leq \epsilon\} \cup \{\lambda \mid 1/\epsilon \leq |\lambda| \leq \infty\} = I^\epsilon \cup I^{1/\epsilon}.$$

The factorization problem (or Riemann-Hilbert problem) with respect to these regions has already been discussed in Chapter 12. It is related to the partial decomposition

$$\Lambda^{\epsilon,1/\epsilon}GL_n\mathbf{C} \supseteq \Lambda_E^1 GL_n\mathbf{C}\, \Lambda_I GL_n\mathbf{C}$$

of the group

$$\Lambda^{\epsilon,1/\epsilon}GL_n\mathbf{C} = \{\gamma : C^\epsilon \cup C^{1/\epsilon} \to GL_n\mathbf{C} \mid \gamma \text{ smooth}\}.$$

Because of the nature of the 2DTL, we use the following modification of this partial decomposition:

$$(\Lambda_{\mathbf{R}}^{\epsilon,1/\epsilon}SL_{n+1}\mathbf{C})_\sigma \supseteq (\Lambda_{E,\mathbf{R}}SL_{n+1}\mathbf{C})_\sigma (\Lambda_{I,\mathbf{R}}^{A,A}SL_{n+1}\mathbf{C})_\sigma.$$

In this (admittedly quite appalling) notation,

(i) A_{n+1} is the subgroup of $SL_{n+1}\mathbf{C}$ consisting of diagonal matrices with positive diagonal entries,

(ii) $\Lambda_I^{A,A}SL_{n+1}\mathbf{C} = \{(\delta^\epsilon, \delta^{1/\epsilon}) \in \Lambda_I SL_{n+1}\mathbf{C} \mid \delta^\epsilon(0) \in A_{n+1}, \ \delta^{1/\epsilon}(\infty) \in A_{n+1}\}$,

(iii) $\mathbf{R}$ indicates that the Reality Assumption is in force, i.e., $\gamma(1/\bar{\lambda}) = \gamma(\lambda)^{-1*}$, and

(iv) σ indicates that we have twisted loops.

With this terminology, our factorization assumption can be stated clearly. It has two components:

(F1) We assume that $F : \mathbf{C} \to (\Lambda SU_{n+1})_\sigma$ is a solution of (Λ_σ) which extends to a map $F : \mathbf{C} \to (\Lambda_{E,\mathbf{R}}SL_{n+1}\mathbf{C})_\sigma$.

(F2) Let $\gamma = (\gamma^\epsilon, \gamma^{1/\epsilon}) \in (\Lambda_{\mathbf{R}}^{\epsilon,1/\epsilon}SL_{n+1}\mathbf{C})_\sigma$, and let $\tilde{F} = \gamma F$. We assume that $\tilde{F} = \tilde{F}_1\tilde{F}_2$, where $\tilde{F}_1, \tilde{F}_2$ take values (respectively) in $(\Lambda_{E,\mathbf{R}}SL_{n+1}\mathbf{C})_\sigma$, $(\Lambda_{I,\mathbf{R}}^{A,A}SL_{n+1}\mathbf{C})_\sigma$.

Remarkably, both (F1) and (F2) are automatically satisfied! This follows from:

Lemma.

(1) For any solution $w_0, \ldots, w_n$ of the (original version of the) 2DTL, there exists a corresponding solution F which takes values in $(\Lambda_{E,\mathbf{R}}SL_{n+1}\mathbf{C})_\sigma$.

(2) $(\Lambda_{\mathbf{R}}^{\epsilon,1/\epsilon}SL_{n+1}\mathbf{C})_\sigma = (\Lambda_{E,\mathbf{R}}SL_{n+1}\mathbf{C})_\sigma (\Lambda_{I,\mathbf{R}}^{A,A}SL_{n+1}\mathbf{C})_\sigma$, i.e., the above partial decomposition is actually a decomposition.

Proof. (1) This is because A_λ, B_λ are well defined for any $\lambda \in \mathbf{C}^*$, and in particular for $\lambda \in E$. (2) The result for un-twisted loops is given at the end of Chapter 12. The necessary modification for twisted loops is proved in the same way, using the twisted Birkhoff and Iwasawa decompositions (given in Theorem 2.3 of Dorfmeister *et al.* [in press]). ■

We reach the following conclusion:

Theorem. *The group* $(\Lambda_{\mathbf{R}}^{\epsilon,1/\epsilon} SL_{n+1}\mathbf{C})_\sigma$ *acts on the set of* $(\Lambda_{E,\mathbf{R}} SL_{n+1}\mathbf{C})_\sigma$- *valued solutions of* (Λ_σ). ∎

This is the "hidden symmetry group" of the 2DTL.

IV. A family of solutions of the 2DTL.

In Chapter 9 we found the following "trivial" solution of the 2DTL:

$$F_0(z,\lambda) = \exp(z\lambda A_0 + \bar{z}\tfrac{1}{\lambda} B_0)$$

where

$$A_0 = \begin{pmatrix} 0 & 0 & 0 & 0 & 1 \\ 1 & 0 & 0 & 0 & 0 \\ 0 & 1 & 0 & 0 & 0 \\ 0 & 0 & 1 & 0 & 0 \\ 0 & 0 & 0 & 1 & 0 \end{pmatrix}, \quad B_0 = \begin{pmatrix} 0 & -1 & 0 & 0 & 0 \\ 0 & 0 & -1 & 0 & 0 \\ 0 & 0 & 0 & -1 & 0 \\ 0 & 0 & 0 & 0 & -1 \\ -1 & 0 & 0 & 0 & 0 \end{pmatrix}.$$

(By analogy with other differential equations of mathematical physics, such a trivial solution is sometimes called a *vacuum solution*.) It is easy to verify that F_0 is a solution (one needs to use the fact that $[A_0, B_0] = 0$), and that F satisfies the Reality Assumption. In fact, F_0 is a $(\Lambda_{E,\mathbf{R}} SL_{n+1}\mathbf{C})_\sigma$-valued solution.

In the last section we showed that the group $(\Lambda_{\mathbf{R}}^{\epsilon,1/\epsilon} SL_{n+1}\mathbf{C})_\sigma$ acts on the set of $(\Lambda_{E,\mathbf{R}} SL_{n+1}\mathbf{C})_\sigma$-valued solutions of (Λ_σ). (This group action is called the *dressing action*.)

Therefore, for every $(\gamma^\epsilon, \gamma^{1/\epsilon}) \in (\Lambda_{\mathbf{R}}^{\epsilon,1/\epsilon} SL_{n+1}\mathbf{C})_\sigma$ we obtain a new $(\Lambda_{E,\mathbf{R}} SL_{n+1}\mathbf{C})_\sigma$-valued solution

$$F = (\gamma^\epsilon, \gamma^{1/\epsilon}) \cdot F_0 = ((\gamma^\epsilon, \gamma^{1/\epsilon}) F_0)_1.$$

(The solution F is said to be obtained by *dressing the vacuum*.) If F also satisfies condition (T), it corresponds to a solution of the 2DTL. This is the family of solutions of the 2DTL that we wish to consider.

We do not claim that *all* solutions of the 2DTL are of this form, of course. Nevertheless, these solutions are in general non-trivial, and we shall investigate them in more detail.

Bibliographical comments.

See the comments at the end of Chapter 15.

Chapter 14: τ-functions and the Bruhat decomposition

I. Explicit solutions of the 1DTL.

In preparation for a more detailed study of the 2DTL, we return to the 1DTL in this section. In Chapter 5, we saw that the (general) solution of the 1DTL is given in terms of a group decomposition

$$SL_n\mathbf{R} = G = G_1 G_2$$

by

$$L(t) = \operatorname{Ad} u(t)^{-1} V$$

where $u(t) = (\exp tV)_1$ and $V = L(0)$. In this chapter we show that there is a *more explicit* formula for the general solution $L(t)$ of the 1DTL.

The groups G_1, G_2 are

$$G_1 = SO_n, \quad G_2 = A_n N'_n = \hat{N}'_n$$

and so we have a factorization

$$\exp tV = S(t) A(t) N(t)$$

where $S(t), A(t), N(t)$ take values (respectively) in SO_n, A_n, N'_n. Thus, $(\exp tV)_1 = S(t)$, $(\exp tV)_2 = A(t) N(t)$, and $A(t)$ is a diagonal matrix whose diagonal entries $d_1(t), \ldots, d_n(t)$ are positive. It turns out that, because of the special form of the matrix $L(t)$, the solution of the 1DTL may also be expressed in terms of the "middle factor" $A(t)$:

Lemma. *In terms of the original variables $q_1, \ldots, q_n$ of the 1DTL, we have*

$$q_i(t) = q_i(0) + \log d_i(t), \quad i = 1, \ldots, n.$$

Proof. We have $S(t) = \exp tV N(t)^{-1} A(t)^{-1}$, so it follows that $L = S^{-1} V S = A N V N^{-1} A^{-1}$. Now, L is of the form

$$\begin{pmatrix} * & Q_{1,2} & & & & \\ Q_{1,2} & * & Q_{2,3} & & & \\ & Q_{2,3} & * & \cdots & & \\ & & \cdots & \cdots & Q_{n-1,n} \\ & & & Q_{n-1,n} & * \end{pmatrix},$$

hence NVN^{-1} is of the form

$$\begin{pmatrix} * & Q_{1,2}(0) & & & & \\ * & * & Q_{2,3}(0) & & & \\ * & * & * & \cdots & & \\ * & * & * & * & Q_{n-1,n}(0) \\ * & * & * & * & * \end{pmatrix},$$

and $ANVN^{-1}A^{-1}$ is of the form

$$\begin{pmatrix} * & d_1 Q_{1,2}(0)d_2^{-1} & & & \\ * & * & d_2 Q_{2,3}(0)d_3^{-1} & & \\ * & * & * & \cdots & \\ * & * & * & * & d_{n-1}Q_{n-1,n}(0)d_n^{-1} \\ * & * & * & * & * \end{pmatrix}.$$

Comparing this with L, we see that

$$Q_{i,i+1} = \frac{d_i}{d_{i+1}} Q_{i,i+1}(0).$$

As $Q_{i,i+1} = e^{q_i - q_{i+1}}$, we obtain $e^{q_i - q_i(0)}/d_i = e^{q_{i+1} - q_{i+1}(0)}/d_{i+1}$. Since $\prod e^{q_i} = \prod d_i = 1$, we must have $e^{q_i - q_i(0)} = ud_i$, where u is an n-th root of unity. Since $d_i(0) = 1$, we must have $u = 1$. ∎

Thanks to this lemma, it suffices to find an explicit formula for $A(t)$. Using elementary matrix theory, we can do this in the following two steps.

Step 1: From Iwasawa to Gauss.

From $\exp tV = S(t)A(t)N(t)$, we have

$$\begin{aligned} (\exp tV)^t(\exp tV) &= N(t)^t A(t)^t S(t)^t S(t) A(t) N(t) \\ &= N(t)^t A(t)^t A(t) N(t) \\ &= N(t)^t A(t)^2 N(t). \end{aligned}$$

Since V is symmetric, we may re-write this formula as

$$\exp 2tV = N(t)^t A(t)^2 N(t).$$

This is the factorization of $\exp 2tV$ with respect to the Gauss decomposition (or, more precisely, partial decomposition)

$$SL_n\mathbf{R} \supseteq N_n A_n N_n'.$$

Step 2: From Gauss to τ-functions.

Surprisingly, it is very easy to find an explicit formula for the "diagonal factor" D in a Gauss decomposition $X = UDL$, where $X \in SL_n\mathbf{R}$, $U \in N_n$, $L \in N_n'$. (We do not assume that the diagonal entries of D are positive, here.) This works because the determinant of a triangular matrix is the product of its diagonal entries, as we shall explain.

Let X_k, U_k, D_k, L_k denote (respectively) the "lower principal $k \times k$ minors" of X, U, D, L. (In other words, X_k is the $k \times k$ matrix $(x_{ij})_{n-k+1 \le i,j \le n}$, if $X = (x_{ij})_{1 \le i,j \le n}$, etc.) Then we have

$$X_k = U_k D_k L_k$$

and therefore $d_{n-k+1} \ldots d_n = \det X_k$. We conclude that the diagonal entries of D are given by

$$d_i = \frac{\det X_{n-i+1}}{\det X_{n-i}}, \quad i = 1, \ldots, n$$

(with the convention that $\det X_0 = 1$). We use the notation $\tau_i(X) = \det X_i$.

Hence we obtain our explicit formula for the solution of the 1DTL.

Theorem.

$$q_i(t) = q_i(0) + \frac{1}{2} \log \frac{\tau^V_{n-i+1}(t)}{\tau^V_{n-i}(t)}, \quad i = 1, \ldots, n$$

where $\tau^V_i(t) = \det(\exp 2tV)_i$. ∎

(When $n = 2$, this agrees with the explicit formula obtained in Chapter 5.)

The functions τ_i, τ^V_i are called *τ-functions*.

The following proposition illustrates the significance of the τ-functions. (In addition, the proof reveals why we use the term "Gauss decomposition"!)

Proposition. *Let* $X \in SL_n\mathbf{R}$. *The Gauss factorization* $X = UDL$ *is possible if and only if* $\tau_1(X), \ldots, \tau_{n-1}(X)$ *are all non-zero.*

Sketch of the proof. It is easy to see that the existence of a Gauss factorization

$$X = LDU$$

is equivalent to the solvability of the linear system

$$Xu = v$$

by "Gaussian elimination, without re-ordering the equations".

For example, consider the following system:

$$
\begin{array}{rcrcrcr}
-x & + & y & + & z & = & 1 \\
x & & & - & z & = & 1 \\
x & - & y & + & z & = & -1.
\end{array}
$$

Let us add the first equation to the second and third equations. We obtain the simpler system

$$
\begin{array}{rcl}
-x \;+\; y \;+\; z &=& 1 \\
y &=& 2 \\
2z &=& 0
\end{array}
$$

from which we can read off the solution $x = 1$, $y = 2$, $z = 0$. By "Gaussian elimination" we mean this method, where the system is transformed (if possible) into triangular form.

In terms of matrices, the transformation can be expressed as

$$
\begin{pmatrix} 1 & 0 & 0 \\ 1 & 1 & 0 \\ 1 & 0 & 1 \end{pmatrix}
\begin{pmatrix} -1 & 1 & 1 \\ 1 & 0 & -1 \\ 1 & -1 & 1 \end{pmatrix}
=
\begin{pmatrix} -1 & 1 & 1 \\ 0 & 1 & 0 \\ 0 & 0 & 2 \end{pmatrix},
$$

or

$$
\begin{pmatrix} -1 & 1 & 1 \\ 1 & 0 & -1 \\ 1 & -1 & 1 \end{pmatrix}
=
\begin{pmatrix} 1 & 0 & 0 \\ -1 & 1 & 0 \\ -1 & 0 & 1 \end{pmatrix}
\begin{pmatrix} -1 & 0 & 0 \\ 0 & 1 & 0 \\ 0 & 0 & 2 \end{pmatrix}
\begin{pmatrix} 1 & -1 & -1 \\ 0 & 1 & 0 \\ 0 & 0 & 1 \end{pmatrix}.
$$

In other words, the calculation produces a Gauss factorization of the original matrix.

If we perform this calculation with the general 3×3 system

$$
\begin{array}{rcl}
a_{11}x \;+\; a_{12}y \;+\; a_{13}z &=& b_1 \\
a_{21}x \;+\; a_{22}y \;+\; a_{23}z &=& b_2 \\
a_{31}x \;+\; a_{32}y \;+\; a_{33}z &=& b_3
\end{array}
$$

then – if $a_{11} \neq 0$ – we obtain the system

$$
\begin{array}{rcl}
x \;+\; \dfrac{a_{12}}{a_{11}}y \;+\; \dfrac{a_{13}}{a_{11}}z &=& \dfrac{b_1}{a_{11}} \\[2ex]
\dfrac{a_{11}a_{22} - a_{12}a_{21}}{a_{11}}y \;+\; \dfrac{a_{11}a_{23} - a_{13}a_{21}}{a_{11}}z &=& \dfrac{a_{11}b_2 - a_{21}}{a_{11}} \\[2ex]
\dfrac{a_{11}a_{32} - a_{12}a_{31}}{a_{11}}y \;+\; \dfrac{a_{11}a_{33} - a_{13}a_{31}}{a_{11}}z &=& \dfrac{a_{11}b_3 - a_{31}}{a_{11}}.
\end{array}
$$

We may perform the final step (to reach triangular form) if $a_{11}a_{22} - a_{12}a_{21} \neq 0$. Thus, in this case, Gauss elimination works if and only if $\tau_1(X), \tau_2(X) \neq 0$. ∎

The above discussion is based on section 4.3 of Perelomov [1990]. The proposition can be found in Chapter 4 of Strang [1976].

II. τ-functions, representations, and Grassmannians.

We sketch now a Lie theoretic proof of the last proposition. The Lie theoretic significance of the (complex) τ-functions

$$\tau_i : SL_n\mathbf{C} \to \mathbf{C}, \ X \mapsto \det X_i, \ i = 1, \ldots, n-1$$

is the key to finding explicit solutions of the 2DTL.

First, we observe that

$$\det X_i = \langle X(e_{n-i+1} \wedge \ldots \wedge e_n), e_{n-i+1} \wedge \ldots \wedge e_n \rangle$$
$$= \langle Xe_{n-i+1} \wedge \ldots \wedge Xe_n, e_{n-i+1} \wedge \ldots \wedge e_n \rangle$$

where $e_1, \ldots, e_n$ is the standard basis of $\mathbf{C}^n$, and where $\langle \, , \, \rangle$ is the standard Hermitian inner product on $\wedge^i \mathbf{C}^n$. Let $\lambda : SL_n\mathbf{C} \to SL_n\mathbf{C}$ be the standard (irreducible) representation of $SL_n\mathbf{C}$ (i.e., λ is the identity map, considered as a representation). In the terminology of representation theory, $e_{n-i+1} \wedge \ldots \wedge e_n$ is a highest weight vector of the (irreducible) representation

$$\wedge^i\lambda : SL_n\mathbf{C} \to SL_{\binom{n}{i}}\mathbf{C}.$$

Next, consider the map

$$X \mapsto [X(e_{n-i+1} \wedge \ldots \wedge e_n)] \in \mathbf{C}P^{\binom{n}{i}-1}$$

(the "projectivized orbit of the highest weight vector"). The isotropy subgroup at the point $[e_{n-i+1} \wedge \ldots \wedge e_n]$ under this action of $SL_n\mathbf{C}$ is

$$P_i = \{X \in SL_n\mathbf{C} \mid X(\mathbf{C}^{n-i})^\perp \subseteq (\mathbf{C}^{n-i})^\perp\}$$

where $(\mathbf{C}^{n-i})^\perp = \mathrm{Span}\{e_{n-i+1}, \ldots, e_n\}$. Hence the orbit of $[e_{n-i+1} \wedge \ldots \wedge e_n]$ (i.e., the image of the above map) is a copy of the Grassmannian $\mathrm{Gr}_i(\mathbf{C}^n) \cong SL_n\mathbf{C}/P_i$. For $X \in SL_n\mathbf{C}$, the element $X(\mathbf{C}^{n-i})^\perp$ of $\mathrm{Gr}_i(\mathbf{C}^n)$ corresponds to the coset XP_i. We use this notation interchangeably from now on.

Finally, let Det denote the (holomorphic) line bundle on $\mathrm{Gr}_i(\mathbf{C}^n)$ whose fibre over the i-plane $\mathrm{Span}\{v_1, \ldots, v_i\}$ is the line $\mathrm{Span}\{v_1 \wedge \ldots \wedge v_i\}$. The total space of Det may be identified with

$$SL_n\mathbf{C} \times_{P_i} \mathrm{Span}\{e_{n-i+1} \wedge \ldots \wedge e_n\}$$

where P_i acts on $\mathrm{Span}\{e_{n-i+1} \wedge \ldots \wedge e_n\}$ via $\wedge^i\lambda$.

We now have all the necessary ingredients for our Lie theoretic description of τ_i:

Proposition.

(1) τ_i defines a holomorphic section of the dual bundle Det*.

(2) For any element $X(\mathbf{C}^{n-i})^\perp$ of $SL_n\mathbf{C}/P_i$, $\tau_i(X(\mathbf{C}^{n-i})^\perp) = 0$ if and only if $X(\mathbf{C}^{n-i})^\perp \notin N_n(\mathbf{C}^{n-i})^\perp$.

Sketch of the proof.

(1) Let G/H be any homogeneous space, and let $\theta : H \to \mathrm{Aut}(V)$ be any representation of H. Then there is a one to one correspondence between sections s of the associated bundle

$$G \times_H V \to G/H$$

and maps

$$S : G \to V$$

such that $S(gh^{-1}) = \theta(h)S(g)$ for all $h \in H, g \in G$. The correspondence is given by $s(gH) = [(g, S(g))]$.

In our case, the map $S : SL_n\mathbf{C} \to \mathrm{Span}\{e_{n-i+1} \wedge \ldots \wedge e_n\}^*$ is given by

$$S(X)(ce_{n-i+1} \wedge \ldots \wedge e_n) = c\langle X(e_{n-i+1} \wedge \ldots \wedge e_n), e_{n-i+1} \wedge \ldots \wedge e_n\rangle.$$

(2) Let $E = X(\mathbf{C}^{n-i})^\perp$. The condition $E \in N_n(\mathbf{C}^{n-i})^\perp$ corresponds to the "Schubert condition"

$$\dim \mathbf{C}^{n-i} \cap E = 0, \ \dim \mathbf{C}^{n-i+1} \cap E = 1, \ \ldots, \ \dim \mathbf{C}^n \cap E = i.$$

This is equivalent to the condition

$$X(e_{n-i+1} \wedge \ldots \wedge e_n) = \text{ non-zero constant } \times e_{n-i+1} \wedge \ldots \wedge e_n$$

(as one sees by writing each $X(e_j)$ in terms of the basis vectors). ∎

To relate this to the proposition of the previous section, we use the fact that

$$P_1 \cap \cdots \cap P_{n-1} = \Delta'_n.$$

Proposition. *Let $X \in SL_n\mathbf{C}$. We have $X \in N_nP_1 \cap \cdots \cap N_nP_{n-1}$ if and only if $X \in N_n\Delta'_n$.* ∎

This can be proved directly.

III. The Bruhat decomposition.

The theory of the previous section has an important geometrical interpretation. Although this geometrical interpretation is not strictly necessary at this point for our study of the 1DTL (or 2DTL), it is related to the Hamiltonian system of Chapter 4, and it will be very useful in our study of harmonic maps in Chapters 16-20. It has various aspects – algebraic, topological, and analytic – and because of its fundamental nature we give a brief description of it in this section.

The algebraic version is based on the (disjoint) decompositions

$$SL_n \mathbf{C} = \bigcup_{s \in \Sigma_n} N_n \, s \, \Delta_n'$$

$$SL_n \mathbf{C} = \bigcup_{s \in \Sigma_n} N_n \, s \, \Delta_n$$

(which may be proved by the well known procedure of reduction of a matrix to "echelon form"). Here, Σ_n denotes the (finite) group of permutation matrices.

These induce (disjoint) decompositions of the flag manifold

$$F_{1,2,\ldots,n}(\mathbf{C}^n) = \bigcup_{s \in \Sigma_n} N_n \, s \, ((\mathbf{C}^{n-1})^\perp \subseteq (\mathbf{C}^{n-2})^\perp \subseteq \ldots \subseteq (\mathbf{C})^\perp \subseteq \mathbf{C}^n)$$

$$F_{1,2,\ldots,n}(\mathbf{C}^n) = \bigcup_{s \in \Sigma_n} N_n \, s \, (\mathbf{C} \subseteq \mathbf{C}^2 \subseteq \ldots \subseteq \mathbf{C}^{n-1} \subseteq \mathbf{C}^n).$$

Each piece here is a "cell", i.e., homeomorphic to $\mathbf{C}^N$ for some N. This fact may be seen by elementary matrix considerations; we give an example in a moment.

Let P_i be the isotropy subgroup at $(\mathbf{C}^{n-i})^\perp$ (under the action of $SL_n \mathbf{C}$ on $\mathrm{Gr}_i(\mathbf{C}^n)$), as in the last section. Since $P_i \supseteq \Delta_n'$, we have

$$SL_n \mathbf{C} = \bigcup_{s \in \Sigma_n} N_n \, s \, P_i \quad \text{and} \quad \mathrm{Gr}_i(\mathbf{C}^n) = \bigcup_{s \in \Sigma_n} N_n \, s \, (\mathbf{C}^{n-i})^\perp.$$

This time, we do not have a disjoint decomposition. However, any two pieces are either disjoint or equal, and the condition to be equal is obviously given by:

$$s \, (\mathbf{C}^{n-i})^\perp = t \, (\mathbf{C}^{n-i})^\perp \iff s^{-1} t \in P_i.$$

It follows that each piece corresponds to a choice of i basis vectors from the standard basis $e_1, \ldots, e_n$. This decomposition of $\mathrm{Gr}_i(\mathbf{C}^n)$ into cells is called the *Schubert decomposition*.

As a very concrete example, let us consider the case $i = 1$, i.e., $\mathrm{Gr}_i(\mathbf{C}^n) = \mathbf{C}P^{n-1}$. In this case we have

$$\mathbf{C}P^{n-1} = \bigcup_{s \in \Sigma_n} N_n \, s \, (\mathbf{C}^{n-1})^{\perp} = \bigcup_{i \in \{1,\dots,n\}} N_n \, s \, V_i$$

where $V_i = \mathbf{C}e_i$. There are n disjoint pieces here, and the i-th piece is homeomorphic to $\mathbf{C}^{i-1}$:

$$N_n V_i = \left\{ \begin{pmatrix} 1 & \cdots & * & * & * & \cdots & * \\ & \ddots & \vdots & \vdots & \vdots & & \vdots \\ & & 1 & * & * & \cdots & * \\ & & & 1 & * & \cdots & * \\ & & & & 1 & \cdots & * \\ & & & & & \ddots & \vdots \\ & & & & & & 1 \end{pmatrix} \begin{pmatrix} 0 \\ \vdots \\ 0 \\ 1 \\ 0 \\ \vdots \\ 0 \end{pmatrix} \right\}$$

$$= \left\{ \begin{pmatrix} 1 & \cdots & 0 & * & 0 & \cdots & 0 \\ & \ddots & \vdots & \vdots & \vdots & & \vdots \\ & & 1 & * & 0 & \cdots & 0 \\ & & & 1 & 0 & \cdots & 0 \\ & & & & 1 & \cdots & 0 \\ & & & & & \ddots & \vdots \\ & & & & & & 1 \end{pmatrix} \begin{pmatrix} 0 \\ \vdots \\ 0 \\ 1 \\ 0 \\ \vdots \\ 0 \end{pmatrix} \right\}.$$

There is one "big cell", namely $N_n V_n \cong \mathbf{C}^{n-1}$. Since the other cells have lower dimension, the big cell is open and dense in $\mathbf{C}P^{n-1}$.

It would not make any difference in the above discussion if we replaced N_n by Δ_n (i.e., we allow arbitrary non-zero diagonal entries). We use N_n because it is more economical.

There is a dual decomposition, in which orbits of N_n are replaced by orbits of N_n'. The dual decomposition of $\mathrm{Gr}_i(\mathbf{C}^n)$ is also called the Schubert decomposition. These two decompositions admit a Morse theoretic interpretation; to describe this we identify $\mathrm{Gr}_i(\mathbf{C}^n)$ with an adjoint orbit M_P in $\mathbf{u}_n$, as in section II of Chapter 4 (see Example (2)). Consider the height function $H^Q : \mathrm{Gr}_i(\mathbf{C}^n) \to \mathbf{R}$, where Q is any element of $\mathbf{u}_n$ with distinct eigenvalues. Then the critical points of H^Q are the points $s\,(\mathbf{C}^{n-i})^{\perp}$ for $s \in \Sigma_n$. The cells $N_n s(\mathbf{C}^{n-i})^{\perp}$ and $N_n' s(\mathbf{C}^{n-i})^{\perp}$ are – not necessarily respectively – the unstable ("descending") and stable ("ascending") manifolds of the critical point $s\,(\mathbf{C}^{n-i})^{\perp}$. (The precise correspondence depends on the choice of Q.)

These decompositions, and the Morse theoretic interpretation, are also valid for any flag manifold $F_{n_1,\dots,n_k}(\mathbf{C}^n)$. In fact, they are valid for any generalized flag manifold G^c/P, where P is a parabolic subgroup of G^c. These more general decompositions (of G^c or G^c/P) are called *Bruhat decompositions*.

Bibliographical comments.

See the comments at the end of Chapter 15.

Chapter 15: Solutions of the two-dimensional Toda lattice

Let $(\gamma^\epsilon, \gamma^{1/\epsilon}) \in (\Lambda_{\mathbf{R}}^{\epsilon, 1/\epsilon} SL_{n+1}\mathbf{C})_\sigma$. Let $F_0(z, \lambda) = \exp(z\lambda A_0 + \bar{z}\lambda^{-1} B_0)$, i.e., the trivial solution of the 2DTL (as in Chapter 10). Consider the factorization

$$(\gamma^\epsilon(\lambda), \gamma^{1/\epsilon}(\lambda)) F_0(z, \lambda) = F(z, \lambda)\,(D(z)N(z, \lambda), D(z)^{-1} N(z, 1/\bar{\lambda})^{-1*})$$

(as in Chapter 13). We have $D(z) \in A_{n+1}$ (the group of diagonal matrices in $SL_{n+1}\mathbf{C}$ with positive diagonal entries), and $N(z, 0) = I$. We know that the factor F is a solution of (Λ_σ). If F also satisfies (T), then from Chapter 13 we know that F corresponds to a "new" solution of the 2DTL. Assuming this to be the case, we consider the problem of finding explicit formulae for the corresponding functions $w_0, \ldots, w_n$.

As in the case of the 1DTL, it turns out that the functions w_i depend only on the "middle factor"

$$D = \begin{pmatrix} d_0 & & & \\ & d_1 & & \\ & & \ddots & \\ & & & d_n \end{pmatrix}.$$

The precise relationship is given by the next lemma.

Lemma. *In terms of the original variables* $w_0, \ldots, w_n$ *of the 2DTL, we have*
$$w_i(z) = -\log d_i(z), \quad i = 0, \ldots, n.$$

Proof. We have

$$F^{-1}F_z = \begin{pmatrix} a_0 & & & \\ & a_1 & & \\ & & \ddots & \\ & & & \ddots \\ & & & & a_n \end{pmatrix} + \begin{pmatrix} 0 & & & & b_0 \\ b_1 & 0 & & & \\ & \ddots & \ddots & & \\ & & \ddots & \ddots & \\ & & & b_n & 0 \end{pmatrix}.$$

We claim that $a_i = (\log d_i^{-1})_z$ and $b_i = u_i d_i^{-1}/d_{i-1}^{-1}$ for some constants u_i. The procedure of Chapter 13 then gives $w_i = \log d_i^{-1}$.

Restricting to $|\lambda| = \epsilon$, we have $F = \gamma^\epsilon F_0 N^{-1} D^{-1}$ and hence

$$
\begin{aligned}
F^{-1}F_z &= (\gamma^\epsilon F_0 N^{-1} D^{-1})^{-1}(\gamma^\epsilon F_0 N^{-1} D^{-1})_z \\
&= DN F_0^{-1}(F_0)_z N^{-1} D^{-1} - DN N_z N^{-1} D^{-1} - D_z D^{-1} \\
&= \lambda DN A_0 N^{-1} D^{-1} - DN N_z N^{-1} D^{-1} - D_z D^{-1}.
\end{aligned}
$$

This remains valid in the region $\{\lambda \mid |\lambda| \leq \epsilon\}$. Using the fact that $N(z, 0) = I$, and evaluating at $\lambda = 0$, we obtain $a_i = -(\log d_i)_z$. As in Chapter 13 we have

$$(b_i)_{\bar{z}} = b_i(\bar{a}_i - \bar{a}_{i-1})$$
$$(\bar{b}_i)_z = \bar{b}_i(a_i - a_{i-1}).$$

From this, and our assumption that b_i is real, we obtain

$$(b_i d_i / d_{i-1})_{\bar{z}} = (b_i d_i / d_{i-1})_z = 0.$$

Hence $b_i = u_i d_{i-1} / d_i$ for some constants u_i, as required. ∎

To find the "middle factor" (in terms of the initial data A_0, B_0), we can use the same method that we used for the 1DTL. *To simplify the exposition, we assume that γ^ϵ extends holomorphically to $\{\lambda \mid \epsilon \leq |\lambda| \leq 1\}$.* In this case, the factorization

$$\gamma^\epsilon F_0 = FDN$$

is simply the Iwasawa factorization of $\gamma^\epsilon F_0|_{S^1}$ (see the end of Chapter 12).

Step 1: From Iwasawa to Birkhoff

Since F satisfies the Reality Assumption, we have $F^\dagger = F^{-1}$, where $F^\dagger(\lambda) = F(1/\bar{\lambda})^*$. From $\gamma^\epsilon F_0 = FDN$, we have

$$(\gamma^\epsilon F_0)^\dagger (\gamma^\epsilon F_0) = N^\dagger D^\dagger F^\dagger FDN = N^\dagger D^2 N.$$

We see that the right hand side is a factorization with respect to the Birkhoff (partial) decomposition

$$(\Lambda SL_{n+1}\mathbf{C})_\sigma \supseteq (\Lambda^1_- SL_{n+1}\mathbf{C})_\sigma A_{n+1} (\Lambda^1_+ SL_{n+1}\mathbf{C})_\sigma.$$

Step 2: From Birkhoff to τ-functions

In the corresponding step for the 1DTL ("From Gauss to τ-functions", Chapter 14), we explained how to find the middle factor in terms of τ-functions. Those τ-functions (for the group $SL_n\mathbf{C}$) depend on the following ingredients:

(1) maximal parabolic subgroups $P_1, \ldots, P_{n-1}$, such that $P_1 \cap \cdots \cap P_{n-1} = \Delta'_n$;

(2) a bundle Det on each Grassmannian $\mathrm{Gr}_i(\mathbf{C}^n) \cong SL_n\mathbf{C}/P_i$;

(3) a holomorphic section of Det^*, whose zero set is the complement of the "big cell" $[N_n] \subseteq SL_n\mathbf{C}/P_i$.

These ingredients exist in the case of any finite dimensional complex semisimple Lie group. *It turns out that the ingredients exist also in the case of the infinite dimensional Lie group* $(\Lambda SL_{n+1}\mathbf{C})_\sigma$! We sketch very briefly the generalization of (1)-(3) above, in the case of the (un-twisted) group $\Lambda SL_{n+1}\mathbf{C}$.

(1) Let

$$P_i = \{\gamma \in \Lambda SL_{n+1}\mathbf{C} \mid \gamma K_i = K_i\}$$

where $K_i = \mathrm{Span}\{e_0, \ldots, e_i\} \oplus \lambda H_+^{(n+1)}$. Thus, $P_n = \Lambda_+ SL_{n+1}\mathbf{C}$. The homogeneous space

$$\Lambda SL_{n+1}\mathbf{C}/P_n = \Lambda SL_{n+1}\mathbf{C}/\Lambda_+ SL_{n+1}\mathbf{C}$$

is a subvariety of

$$\Lambda GL_{n+1}\mathbf{C}/\Lambda_+ GL_{n+1}\mathbf{C} \cong \mathrm{Gr}^{(n+1)}.$$

We have

$$\Lambda SL_{n+1}\mathbf{C}/P_i \cong \Lambda SL_{n+1}\mathbf{C}/P_n$$

for $i = 0, \ldots, n$. The fact that

$$P_0 \cap \cdots \cap P_n = \Lambda_+^\Delta SL_{n+1}\mathbf{C}$$

has already been noted in Chapter 12.

(2) There exists a holomorphic line bundle Det on the Grassmannian $\mathrm{Gr}^{(n+1)}$, which is analogous to the bundle Det on $\mathrm{Gr}_i(\mathbf{C}^n)$.

(3) There exists a holomorphic section of Det*, whose zero set is the complement of the "big cell" $[\Lambda_-^{N'} GL_{n+1}\mathbf{C}] \subseteq \Lambda GL_{n+1}\mathbf{C}/\Lambda_+ GL_{n+1}\mathbf{C}$. (A similar statement holds for the restriction of Det to $\Lambda SL_{n+1}\mathbf{C}/\Lambda_+ SL_{n+1}\mathbf{C}$.)

A full discussion of (2) and (3) is given in Pressley and Segal [1986]. (A brief summary is also given in Segal and Wilson [1985].)

Let $\gamma \in \Lambda SL_{n+1}\mathbf{C}$. From (3) we obtain a condition for the existence of a factorization $\gamma = \gamma_1\gamma_2$ where $\gamma_1 \in \Lambda_-^{N'} SL_{n+1}\mathbf{C}$, $\gamma_2 \in \Lambda_+^\Delta SL_{n+1}\mathbf{C}$. It can be shown that the "middle factor" of γ_2 is given in terms of an "infinite determinant". It follows from this that each d_i can be expressed in terms of an infinite determinant. This leads to an explicit formula for our solutions of the 2DTL.

Bibliographical comments for Chapters 13-15.

A more detailed loop theoretic treatment of the 2DTL can be found in McIntosh [1994a; 1994b]. We have presented a simplified version of this theory.

A good reference for the Schubert decomposition of a Grassmannian, and its role in topology, is Milnor and Stasheff [1974]. For the Morse theoretic interpretation of this decomposition, see Parker [1972]. Just as the Schubert decomposition of a Grassmannian is a special case of the Bruhat decomposition of G^c/P, the theory of Chapter 14, section II, generalizes to the case of any complex semisimple Lie group G^c. The subgroups P_i must be replaced by the maximal parabolic subgroups of G, and the representations $\wedge^i\lambda$ by the fundamental irreducible representations. This is part of "Borel-Weil theory", a brief summary of which is given in Segal [1982]; Carter *et al.* [1995].

The τ-functions for the 1DTL are a manifestation of the deep relationship which exists between Lax equations and representation theory of finite dimensional Lie groups. Further information on this theory, which is due to Kostant, may be found in Kostant [1979]; Goodman and Wallach [1984b].

The generalization of τ-functions to the case of infinite dimensional Lie groups has played an important role in other integrable systems. Some references for this are Kashiwara and Miwa [1981]; Date *et al.* [1981-2]; Wilson [1985]; Segal and Wilson [1985].

Chapter 16: Harmonic maps from C to a Lie group

I. The loop group formulation of the harmonic map equation.

Recall from Chapter 9 that a map $\phi : \mathbf{C} \to G$ is *harmonic* if and only if

$$(\phi^{-1}\phi_{\bar{z}})_z + (\phi^{-1}\phi_z)_{\bar{z}} = 0.$$

Here, G is a compact Lie group. We assume that G is a matrix group, as usual.

In addition, from Chapter 9, we know that the harmonic map equation is equivalent to the zero-curvature equation

$$(A_\lambda)_{\bar{z}} - (B_\lambda)_z = [A_\lambda, B_\lambda]$$

where $A_\lambda = \frac{1}{2}(1 - \lambda^{-1})A$, $B_\lambda = \frac{1}{2}(1 - \lambda)B$.

Because of the geometrical interpretation of zero-curvature equations, we know that this is equivalent to the system

(Ω)
$$F^{-1}F_z = \tfrac{1}{2}(1 - \tfrac{1}{\lambda})A$$
$$F^{-1}F_{\bar{z}} = \tfrac{1}{2}(1 - \lambda)B$$

where $F : \mathbf{C} \to \Omega G$. Here we have $A, B : \mathbf{C} \to \mathbf{g} \otimes \mathbf{C}$, with $A = c(B)$, where $c : \mathbf{g} \otimes \mathbf{C} \to \mathbf{g} \otimes \mathbf{C}$ is "conjugation with respect to the real form $\mathbf{g}$", as in section I of Chapter 9. (If $G = U_n$, then $c(X) = -X^*$.) The second equation here is equivalent to the first equation: It may be obtained from the first equation by applying the involution c.

Let us repeat our justification for saying that (Ω) is equivalent to the harmonic map equation, as this is very important. Given maps F, A, B which satisfy (Ω), then the map $\phi(z) = F(z, -1)$ is harmonic (and we necessarily have $A = \phi^{-1}\phi_z, B = \phi^{-1}\phi_{\bar{z}}$). Conversely, given a harmonic map ϕ, there exist maps F, A, B which satisfy (Ω) with $\phi(z) = F(z, -1)$ (and we necessarily have $A = \phi^{-1}\phi_z, B = \phi^{-1}\phi_{\bar{z}}$). This F is unique up to multiplication on the left by a loop $\gamma \in \Omega G$ such that $\gamma(-1) = e$.

This formulation of the harmonic map equation was first treated systematically in Uhlenbeck [1989]. Following Uhlenbeck, we sometimes call a solution F of (Ω) an *extended solution*, or an *extended harmonic map*.

The above three versions of the harmonic map equation are analogous to the three versions of the 2DTL in Chapter 13. As in the case of the 2DTL, it is the last version, (Ω), which we use, because this version reveals the role of the loop group.

It is useful to reformulate (Ω) as follows:

(Ω)
$$F^{-1}F_z = \text{linear in } \tfrac{1}{\lambda} \ (= U + \tfrac{1}{\lambda}V \text{ for some } U, V)$$
$$F^{-1}F_{\bar{z}} = \text{linear in } \lambda \ (= X + \lambda Y \text{ for some } X, Y)$$

where $F : \mathbf{C} \to \Omega G$ and $U, V, X, Y : \mathbf{C} \to \mathbf{g} \otimes \mathbf{C}$, with $U = c(X), Y = c(V)$. This seems at first sight to be a weaker condition, but it isn't. The fact that F takes values in ΩG implies that $F^{-1}F_z$ and $F^{-1}F_{\bar{z}}$ take values in $\Omega\mathbf{g}$. Hence we must have $U + V = 0$, $X + Y = 0$.

A slightly more subtle reformulation is:

$$(\Lambda) \qquad \begin{aligned} F^{-1}F_z &= \text{linear in } \tfrac{1}{\lambda} \; (= U + \tfrac{1}{\lambda}V \text{ for some } U, V) \\ F^{-1}F_{\bar{z}} &= \text{linear in } \lambda \; (= X + \lambda Y \text{ for some } X, Y) \end{aligned}$$

where $F : \mathbf{C} \to \Lambda G$. In other words, we allow F to take values in ΛG, rather than ΩG. We claim that (Λ) is equivalent to (Ω). It is obvious that a solution of (Ω) is a solution of (Λ). Conversely, if F is a solution of (Λ), then we claim that

$$H(z, \lambda) = F(z, \lambda)F(z, 1)^{-1}$$

is a solution of (Ω). To prove this, let us write $h(z) = F(z, 1)^{-1}$. Then we have

$$\begin{aligned} H^{-1}H_z &= h^{-1}F^{-1}(Fh)_z \\ &= h^{-1}F^{-1}(F_z h + Fh_z) \\ &= h^{-1}F^{-1}F_z h + h^{-1}h_z \\ &= (h^{-1}Uh + h^{-1}h_z) + \tfrac{1}{\lambda}(h^{-1}Vh). \end{aligned}$$

Thus, H is another solution of (Λ). Since $H(z, 1) = 1$, H must actually be a solution of (Ω).

This argument says that if $F : \mathbf{C} \to \Lambda G$ is a solution of (Λ), then $[F] : \mathbf{C} \to \Lambda G/G \cong \Omega G$ is a solution of (Ω).

One advantage of (Λ) over (Ω) is that it reveals an action of S^1: If $u \in S^1$, and F is a solution of (Λ), then $(u \cdot F)(z, \lambda) = F(z, u\lambda)$ is obviously a solution of (Λ) as well. This action of S^1 can be transferred to (Ω), but it is given there by the less natural formula $(u \cdot F)(z, \lambda) = F(z, u\lambda)F(z, u)^{-1}$.

On the other hand, passing from (Ω) to (Λ) introduces further "gauge ambiguity". If F is a solution of (Ω), then γF is a solution which represents essentially the same harmonic map, for any $\gamma \in \Omega G$. But if F is a solution of (Λ), then γFh represents essentially the same harmonic map, for any $\gamma \in \Lambda G$ and any (smooth) $h : \mathbf{C} \to G$.

II. The symmetry group of the harmonic map equation.

Let F be a solution of (Λ). As in the case of the 2DTL, we may assume that $F(z, \lambda)$ extends holomorphically in λ to the region $\mathbf{C}^*$, because the equation (Λ) is defined for all $\lambda \in \mathbf{C}^*$. To be more precise: We may assume that we have $F : \mathbf{C} \times \mathbf{C}^* \to G^c$, holomorphic in the second variable, such that $F(z, 1/\bar{\lambda}) = C(F(z, \lambda))$, where $C : G^c \to G^c$ is the involution corresponding to $c : \mathbf{g} \otimes \mathbf{C} \to \mathbf{g} \otimes \mathbf{C}$. (If $G = U_n$, we have $G^c = GL_n\mathbf{C}$, and $C(A) = A^{-1*}$, $c(X) = -X^*$.)

In particular, we may assume that

$$F : \mathbf{C} \to \Lambda_{E,\mathbf{R}} G^c,$$

where $E = \{\lambda \mid \epsilon \le |\lambda| \le 1/\epsilon\}$ (as in Chapter 12), and where "**R**" indicates that we make the Reality Assumption $\gamma(1/\bar{\lambda}) = C(\gamma(\lambda))$.

Let $(\gamma^\epsilon, \gamma^{1/\epsilon})$ be any element of $\Lambda_{\mathbf{R}}^{\epsilon,1/\epsilon} G^c$. We have the decompositon

$$\Lambda_{\mathbf{R}}^{\epsilon,1/\epsilon} G^c = \Lambda_{E,\mathbf{R}} G^c \, \Lambda_{I,\mathbf{R}}^{\hat{N},\hat{N}'} G^c$$

as in Chapter 12. (There we considered the case $G^c = GL_{n+1}\mathbf{C}$, but the general case is equally valid.) Hence we may define

$$\tilde{F} = (\gamma^\epsilon, \gamma^{1/\epsilon}) \cdot F = ((\gamma^\epsilon, \gamma^{1/\epsilon})F)_E.$$

Proposition. *$\tilde{F}$ is a solution of* (Λ).

Proof. We use the argument of Chapter 10, exactly as we did in the case of the 2DTL in Chapter 13. Since this is such a fundamental result, however, we shall repeat the argument here.

First, by definition, $\tilde{F}$ extends holomorphically in λ to the region E. Hence the same is true of $\tilde{F}^{-1}\tilde{F}_z$.

Next, on re-writing $\tilde{F}$ as $\tilde{F} = (\gamma^\epsilon, \gamma^{1/\epsilon})FH^{-1}$, where $H = ((\gamma^\epsilon, \gamma^{1/\epsilon})F)_I$, we have

$$\tilde{F}^{-1}\tilde{F}_z = HF^{-1}(\gamma^\epsilon, \gamma^{1/\epsilon})^{-1}((\gamma^\epsilon, \gamma^{1/\epsilon})F_z H^{-1} - (\gamma^\epsilon, \gamma^{1/\epsilon})FH^{-1}H_z H^{-1})$$

$$= H^{-1}F^{-1}F_z H^{-1} - H_z H^{-1}$$

$$= H^{-1}(A + \tfrac{1}{\lambda}B)H^{-1} - H_z H^{-1}.$$

This extends holomorphically in λ to the region $I - \{0\}$, with a simple pole at $\lambda = 0$.

It follows from this and elementary complex analysis that $\tilde{F}^{-1}\tilde{F}_z$ extends holomorphically in λ to $\mathbf{C}^* \cup \infty$, with a simple pole at $\lambda = 0$. This means that it is linear in $1/\lambda$.

By the Reality Assumption (or by a similar argument for $\tilde{F}^{-1}\tilde{F}_{\bar{z}}$), we see that $\tilde{F}^{-1}\tilde{F}_z$ is linear in λ. Hence $\tilde{F}$ satisfies (Λ). ∎

Thus, as in the case of the 2DTL, we have found a symmetry group of the harmonic map equation, namely $\Lambda_{\mathbf{R}}^{\epsilon,1/\epsilon} G^c$:

Theorem. *The group* $\Lambda_{\mathbf{R}}^{\epsilon,1/\epsilon} G^c$ *acts on the set of* $\Lambda_{E,\mathbf{R}} G^c$-*valued extended harmonic maps.* ∎

It should be noted that the action of the subgroup $\Lambda_{I,\mathbf{R}}^{\hat{N},\hat{N}'} G^c$ is the main point here. The subgroup $\Lambda_{E,\mathbf{R}} G^c$ acts simply by multiplication, and this corresponds to translation of a harmonic map by a (constant) element of G.

Remark: If we wish to use (Ω) instead of (Λ), then we must use the decomposition

$$\Lambda_{\mathbf{R}}^{\epsilon,1/\epsilon} G^c = \Lambda_{E,\mathbf{R}}^1 G^c \, \Lambda_{I,\mathbf{R}} G^c$$

instead of the above decomposition. This change does not seriously affect the theory, however.

III. Some families of harmonic maps.

By applying the action of the symmetry group to an obvious solution, we obtain a family of (less obvious) solutions. We shall now consider two such examples.

Examples:

(1) From Chapter 10 we have the following obvious solution of (Ω):

$$F_0(z,\lambda) = \exp(\tfrac{1}{2}z(1 - \tfrac{1}{\lambda})A_0 + \tfrac{1}{2}\bar{z}(1 - \lambda)B_0)$$

where $A_0, B_0 \in \mathbf{g} \otimes \mathbf{C}$ and $B_0 = c(A_0)$, $[A_0, B_0] = 0$. The corresponding harmonic map ϕ is given by $\phi(z) = \exp(zA_0 + \bar{z}B_0) = \exp zA_0 \exp \bar{z}B_0$.

Similarly, the map

$$F_0(z,\lambda) = \exp(z\tfrac{1}{\lambda}A_0 + \bar{z}\lambda B_0)$$

is an obvious solution of (Λ).

(2) The following solution is perhaps not obvious, but it is very simple, and it arises naturally from the differential geometric theory of harmonic maps. It is

$$\phi = i \circ f$$

where

$$f : \mathbf{C} \to G/K$$

is a holomorphic map into a compact Hermitian symmetric space G/K, and where

$$i : G/K \to G$$

is the "Cartan immersion". (The Cartan immersion of a symmetric space will be discussed in Chapter 18.)

It is well known (see Eells and Lemaire [1978]) that a holomorphic map $u : M \to N$ between Kähler manifolds is harmonic, and that the composition of a harmonic map $u : M \to N$ with a totally geodesic map $v : N \to P$ is harmonic. Hence, from these general principles, our map $\phi = i \circ f$ is harmonic. We give a direct proof of this, for the special case where $G = U_n$, $G/K = \mathrm{Gr}_k(\mathbf{C}^n)$. Let $f : \mathbf{C} \to \mathrm{Gr}_k(\mathbf{C}^n)$ be holomorphic. We have an embedding

$$i : \mathrm{Gr}_k(\mathbf{C}^n) \to U_n, \quad V \mapsto \pi_V - \pi_{V^\perp}$$

where $\pi_V : \mathbf{C}^n \to \mathbf{C}^n$ denotes the Hermitian projection operator onto the k-plane V. (This is essentially the Cartan immersion, which is an embedding in this case.) Let us consider

$$F : \mathbf{C} \to \Omega U_n, \quad F(z, \lambda) = \pi_{f(z)} + \lambda \pi_{f(z)^\perp}.$$

An elementary calculation (see below) shows that F is a solution of (Ω). The harmonic map $\phi(z) = i(f(z)) = F(z, -1)$ is simply the composition

$$\mathbf{C} \xrightarrow{\ f\ } \mathrm{Gr}_k(\mathbf{C}^n) \xrightarrow{\ i\ } U_n,$$

as required

Example (2) suggests the question: Is there a loop theoretic version of the harmonic map equation, for maps $\mathbf{C} \to G/K$? The answer is yes, and we consider this theory in Chapter 18. In particular, we shall explain Example (2) from this point of view.

[Here is the calculation referred to above. We have

$$F(\ , \lambda)^{-1} F_z(\ , \lambda) = (\pi_f + \tfrac{1}{\lambda}\pi_{f^\perp})\tfrac{\partial}{\partial z}(\pi_f + \lambda \pi_{f^\perp})$$

$$= (\pi_f + \tfrac{1}{\lambda}\pi_{f^\perp})[\tfrac{\partial}{\partial z} \circ (\pi_f + \lambda \pi_{f^\perp}) - (\pi_f + \lambda \pi_{f^\perp}) \circ \tfrac{\partial}{\partial z}]$$

by making use of the identity

$$\tfrac{\partial}{\partial z}X = \tfrac{\partial}{\partial z} \circ X - X \circ \tfrac{\partial}{\partial z}$$

for $X : \mathbf{C} \to M_n\mathbf{C}$, where $\tfrac{\partial}{\partial z}, X$ on the right hand side are considered as operators on functions $s : \mathbf{C} \to \mathbf{C}^n$. Thus,

$$F(\ , \lambda)^{-1} F_z(\ , \lambda) = (\pi_f + \tfrac{1}{\lambda}\pi_{f^\perp}) \circ \tfrac{\partial}{\partial z} \circ (\pi_f + \lambda \pi_{f^\perp}) - \tfrac{\partial}{\partial z}$$

$$= (\tfrac{1}{\lambda} - 1)\pi_{f^\perp} \circ \tfrac{\partial}{\partial z} \circ \pi_f + (\lambda - 1)\pi_f \circ \tfrac{\partial}{\partial z} \circ \pi_{f^\perp}.$$

We claim that $\pi_f \circ \tfrac{\partial}{\partial z} \circ \pi_{f^\perp} = 0$. To see this, we use the well known identification

$$T\,\mathrm{Gr}_k(\mathbf{C}^n) \otimes \mathbf{C} = T_{1,0}\,\mathrm{Gr}_k(\mathbf{C}^n) \oplus T_{0,1}\,\mathrm{Gr}_k(\mathbf{C}^n)$$

$$= \mathrm{Hom}(H, H^\perp) \oplus \mathrm{Hom}(H^\perp, H)$$

where H is the tautological vector bundle (of rank k) on $\text{Gr}_k(\mathbf{C}^n)$. Since f is holomorphic, the $(0,1)$-part of $Df(\frac{\partial}{\partial z})$ is zero. Now, in the above identification, we have

$$Df(\tfrac{\partial}{\partial z}) = \pi_{f^\perp} \circ \tfrac{\partial}{\partial z} \circ \pi_f \oplus \pi_f \circ \tfrac{\partial}{\partial z} \circ \pi_{f^\perp}$$

so we conclude that $\pi_f \circ \frac{\partial}{\partial z} \circ \pi_{f^\perp} = 0$, as required.]

IV. Example: An orbit of the action.

Let's consider in detail the orbit under $\Lambda_{\mathbf{R}}^{\epsilon,1/\epsilon} GL_n\mathbf{C}$ of the extended solution $\pi_f + \lambda\pi_{f^\perp}$ (from Example (2) above). As we explained in section II, it suffices to consider the orbit of the group $\Lambda_{I,\mathbf{R}} GL_n\mathbf{C}$.

First, for the special case of a constant loop, i.e., $(\gamma^\epsilon, \gamma^{1/\epsilon}) = (X, X^{-1*})$, $X \in GL_n\mathbf{C}$, it is easy to compute the action:

Proposition. $(X, X^{-1*}) \cdot (\pi_f + \lambda\pi_{f^\perp}) = \pi_{Xf} + \lambda\pi_{(Xf)^\perp}$.

Proof. Let $(X, X^{-1*}) \cdot (\pi_f + \lambda\pi_{f^\perp}) = F_1 F_2$, where F_1, F_2 take values in $\Lambda_{E,\mathbf{R}} GL_n\mathbf{C}$, $\Lambda_{I,\mathbf{R}} GL_n\mathbf{C}$ respectively. We have to show that $F_1 = \pi_{Xf} + \lambda\pi_{(Xf)^\perp}$. To do this, it suffices to show that $(\pi_{Xf} + \lambda\pi_{(Xf)^\perp})^{-1} X(\pi_f + \lambda\pi_{f^\perp})$ takes values in $\Lambda_{I,\mathbf{R}} GL_n\mathbf{C}$. And to do this, it suffices to verify that the value of this expression at $\lambda = 0$ is an invertible matrix. The value is just $\pi_{Xf} X \pi_f + \pi_{(Xf)^\perp} X \pi_{f^\perp}$, which is indeed in $GL_n\mathbf{C}$. ∎

We claim next that the action of the whole group $\Lambda_{I,\mathbf{R}} GL_n\mathbf{C}$ "collapses" to the above action of $GL_n\mathbf{C}$. Let $\Lambda'_{I,\mathbf{R}} GL_n\mathbf{C}$ be the subgroup of $\Lambda_{I,\mathbf{R}} GL_n\mathbf{C}$ whose elements are of the form

$$(\gamma(\lambda), \gamma(1/\bar{\lambda})^{-1*}), \quad \text{where } \gamma(\lambda) = \sum_{i\geq 0} A_i\lambda^i \text{ with } A_0 = I.$$

The quotient group $\Lambda_{I,\mathbf{R}} GL_n\mathbf{C}/\Lambda'_{I,\mathbf{R}} GL_n\mathbf{C}$ is isomorphic to $GL_n\mathbf{C}$. We justify our claim by showing that the group $\Lambda'_{I,\mathbf{R}} GL_n\mathbf{C}$ acts trivially:

Theorem (Uhlenbeck [1989]; see also §2 of Guest and Ohnita [1993]). *If* $(\gamma^\epsilon, \gamma^{1/\epsilon}) \in \Lambda'_{I,\mathbf{R}} GL_n\mathbf{C}$, *then we have* $(\gamma^\epsilon, \gamma^{1/\epsilon}) \cdot (\pi_f + \lambda\pi_{f^\perp}) = \pi_f + \lambda\pi_{f^\perp}$.

Proof. First we derive an explicit integral formula for $(gh)_E$, where $g \in \Lambda_{\mathbf{R}}^{\epsilon,1/\epsilon} GL_n\mathbf{C}$, $h \in \Lambda_{E,\mathbf{R}}^1 GL_n\mathbf{C}$. For a fixed $\lambda \in E$, we integrate the following function over the contour $C^\epsilon \cup C^{1/\epsilon}$. (The circle C^ϵ is given the clockwise orientation, and $C^{1/\epsilon}$ the opposite orientation.)

$$f(\mu) = \frac{((h^{-1}gh)^{-1}(\mu) - I)(h^{-1}gh)_E(\mu)}{(\mu - 1)(\mu - \lambda)}$$

$$= \frac{(h^{-1}gh)_I^{-1}(\mu) - (h^{-1}gh)_E(\mu)}{(\mu - 1)(\mu - \lambda)}$$

$$= \frac{(h^{-1}gh)_I^{-1}(\mu)}{(\mu - 1)(\mu - \lambda)} - \frac{h^{-1}gh)_E(\mu)}{\lambda - 1}\left\{\frac{1}{\mu - \lambda} - \frac{1}{\mu - 1}\right\}.$$

Integrating, and using the Cauchy integral formula, we obtain

$$\frac{1}{2\pi\sqrt{-1}}\int_{C^\epsilon \cup C^{1/\epsilon}} f(\mu)d\mu = -\frac{(h^{-1}gh)_E(\lambda) - (h^{-1}gh)_E(1)}{\lambda - 1}$$

$$= \frac{(h^{-1}gh)_E(\lambda) - I}{1 - \lambda}$$

$$= \frac{h^{-1}(\lambda)(gh)_E(\lambda) - I}{1 - \lambda}.$$

So our formula is

$$(gh)_E(\lambda) - h(\lambda) = \frac{h(\lambda)(1 - \lambda)}{2\pi\sqrt{-1}}\int_{C^\epsilon \cup C^{1/\epsilon}}\frac{((h^{-1}gh)^{-1}(\mu) - I)(h^{-1}gh)_E(\mu)}{(\mu - 1)(\mu - \lambda)}d\mu.$$

If $\{g_t\}$ is a curve in $\Lambda'_{I,\mathbf{R}}GL_n\mathbf{C}$, with $g_0 = I$, then differentiation of the formula gives

$$\frac{d}{dt}(g_t h)_E|_{t=0} = \frac{(\lambda - 1)h(\lambda)}{2\pi\sqrt{-1}}\int_{C^\epsilon \cup C^{1/\epsilon}}\frac{h^{-1}(\mu)\frac{d}{dt}g_t|_{t=0}h(\mu)}{(\mu - 1)(\mu - \lambda)}d\mu.$$

If we take

$$h(\lambda) = \pi_f + \lambda\pi_{f\perp}$$

$$\tfrac{d}{dt}g_t|_{t=0} = \sum_{i\geq 1}A_i\lambda^i \ (\text{on } C^\epsilon)$$

then $\frac{d}{dt}(g_t h)_E|_{t=0}$ is zero, because the numerator in the integrand extends holomorphically to the regions $\{\lambda \mid |\lambda| < \epsilon\}$ and $\{\lambda \mid |\lambda| > 1/\epsilon\}$. It follows that the "infinitesimal" action of $\Lambda'_{I,\mathbf{R}}GL_n\mathbf{C}$ fixes $\pi_f + \lambda\pi_{f\perp}$. Since the group $\Lambda'_{I,\mathbf{R}}GL_n\mathbf{C}$ is a connected Lie group, it must act trivially as well. ∎

We conclude that the orbit of $\pi_f + \lambda\pi_{f\perp}$ under $\Lambda_{I,\mathbf{R}}GL_n\mathbf{C}$ is the same as the orbit of $\pi_f + \lambda\pi_{f\perp}$ under the much smaller group $GL_n\mathbf{C}$. The action of the latter group is induced by the natural action of $GL_n\mathbf{C}$ on $\mathrm{Gr}_k(\mathbf{C}^n)$. It turns out that this behaviour is typical of the case when the harmonic map extends from $\mathbf{C}$ to $\mathbf{C}\cup\infty$. For more general harmonic maps, however, the orbits of $\Lambda_{I,\mathbf{R}}GL_n\mathbf{C}$ are usually infinite dimensional.

Bibliographical comments.

See the comments at the end of Chapter 17.

Chapter 17: Harmonic maps from C to a Lie group (continued)

In this chapter we describe an alternative action of the group $\Lambda_{\mathbf{R}}^{\epsilon,1/\epsilon} G^c$ on harmonic maps $\phi : \mathbf{C} \to G$, following Guest and Ohnita [1993].

There are two restrictions on our discussion. First, we assume that

$$G = U_n, \quad G^c = GL_n\mathbf{C}.$$

(The case of general G can be treated by embedding G in some suitable U_n.) Second, and this is a more serious matter, it is known only that the action agrees with the action of Chapter 16 if ϕ extends to the compactification $S^2 = \mathbf{C} \cup \infty$ of $\mathbf{C}$ (or, more generally, if ϕ has "finite uniton number", a term we explain in due course). Subject to these restrictions, the new action has considerable technical advantages over the old action.

I. A new symmetry group for the harmonic map equation.

As motivation, we reconsider the harmonic map $i \circ f$ of Chapter 16, where

$$f : \mathbf{C} \to \mathrm{Gr}_k(\mathbf{C}^n)$$

is holomorphic, and

$$i : \mathrm{Gr}_k(\mathbf{C}^n) \to U_n$$

is given by $i(V) = \pi_V - \pi_V^{\perp}$. This corresponds to the solution

$$F(z, \lambda) = \pi_{f(z)} + \lambda \pi_{f(z)}^{\perp}$$

of (Ω). We showed in section IV of Chapter 16 that the orbit of F under $\Lambda_{\mathbf{R}}^{\epsilon,1/\epsilon} GL_n\mathbf{C}$ is given simply by the orbit of f under $GL_n\mathbf{C}$, where $GL_n\mathbf{C}$ acts on $\mathrm{Gr}_k(\mathbf{C}^n)$ in the natural way.

The new action is a generalization of this example. We consider a solution $F : \mathbf{C} \to \Omega U_n$ of (Ω), then identify ΩU_n with $\mathrm{Gr}^{(n)}$ as in Chapter 12, and finally apply the natural action of $\Lambda GL_n\mathbf{C}$ on $\mathrm{Gr}^{(n)}$. We shall see that this new action agrees with the old action, at least when F corresponds to a harmonic map $S^2 \to U_n$.

First we reformulate our system of equations (Ω) in terms of $\mathrm{Gr}^{(n)}$, in a way that was first pointed out in Segal [1989]. Let $W : \mathbf{C} \to \mathrm{Gr}^{(n)}$ correspond to $F : \mathbf{C} \to \Omega U_n$, under the identification $\Omega U_n \cong \mathrm{Gr}^{(n)}$. Thus, $W(z) = F(z,)H_+^{(n)}$. Consider the following conditions:

(Gr)
$$W_z \subseteq \tfrac{1}{\lambda} W$$
$$W_{\bar{z}} \subseteq W.$$

The first condition means that the vector $\partial s(z)/\partial z$ is contained in the subspace $\lambda^{-1}W(z)$ of $H^{(n)}$, for every (smooth) map $s : \mathbf{C} \to H^{(n)}$ such that $s(z) \in W(z)$. The second condition is interpreted in a similar way. By the calculation at the end of section III of Chapter 16, the second condition says that $W : \mathbf{C} \to \mathrm{Gr}^{(n)}$ is *holomorphic*.

Proposition. *F is a solution of* (Ω) *if and only if W is a solution of* (Gr).

Proof. If F is a solution of (Ω), then it is obvious that W is a solution of (Gr). Conversely, assume that $W = FH^{(n)}_+$ satisfies (Gr). Then we have

$$F^{-1}F_z H^{(n)}_+ \subseteq \tfrac{1}{\lambda}H^{(n)}_+, \quad F^{-1}F_{\bar z}H^{(n)}_+ \subseteq H^{(n)}_+.$$

From the first of these, we have $F^{-1}F_z = \sum_{i \geq -1} A_i\lambda^i$. From the second, we have $F^{-1}F_{\bar z} = \sum_{i \geq 0} B_i\lambda^i$. On applying the transformation $X \mapsto -X^*$, we obtain $F^{-1}F_{\bar z} = \sum_{i \geq -1} C_i\lambda^{-i}$ and $F^{-1}F_z = \sum_{i \geq 0} D_i\lambda^{-i}$. Combining all four expressions, we find that $F^{-1}F_z$ is linear in λ^{-1}, and $F^{-1}F_{\bar z}$ is linear in λ, as required. ∎

For any $\gamma \in \Lambda GL_n\mathbf{C}$ and any solution $W(= FH^{(n)}_+)$ of the system (Gr), we define
$$\gamma \cdot W = \gamma W \; (= \gamma FH^{(n)}_+),$$
i.e., we use the natural action of $\Lambda GL_n\mathbf{C}$ on $\mathrm{Gr}^{(n)} \cong \Lambda GL_n\mathbf{C}/\Lambda_+ GL_n\mathbf{C}$.

Via the identification $\Omega U_n \cong \mathrm{Gr}^{(n)}$, this action of $\Lambda GL_n\mathbf{C}$ on ΩU_n is given simply by $\gamma\cdot\delta = (\gamma\delta)_u$, where the factorization $\gamma\delta = (\gamma\delta)_u(\gamma\delta)_+$ is the factorization with respect to the decomposition $\Lambda GL_n\mathbf{C} = \Omega U_n\Lambda_+ GL_n\mathbf{C}$. Hence, $\gamma W : \mathbf{C} \to \mathrm{Gr}^{(n)}$ corresponds to the map $(\gamma F)_u : \mathbf{C} \to \Omega U_n$. We could therefore define our action in the following equivalent way:

$$\gamma \cdot F = (\gamma F)_u.$$

However, to avoid confusion, we generally express the old action in terms of F, and the new action in terms of W.

Proposition. *$\gamma \cdot W$ is also a solution of* (Gr).

Proof. The proof is trivial: $\Lambda GL_n\mathbf{C}$ acts linearly on $H^{(n)}_+$, and commutes with multiplication by λ^{-1}! ∎

Let us now consider the relationship between the new action of $\Lambda GL_n\mathbf{C}$ and the old action of $\Lambda_{\mathbf{R}}^{\epsilon,1/\epsilon}GL_n\mathbf{C}$. We begin with some preliminary remarks.

(1) For the old action, we must use the decomposition $\Lambda^{\epsilon,1/\epsilon}GL_n\mathbf{C} = \Lambda^1_{E,\mathbf{R}}GL_n\mathbf{C}\,\Lambda_{I,\mathbf{R}}GL_n\mathbf{C}$ (because we use (Ω) rather than (Λ)).

(2) As pointed out in Chapter 16, it suffices to consider the action of $\Lambda_{I,\mathbf{R}}GL_n\mathbf{C}$.

(3) Let $\gamma \in \Lambda_+GL_n\mathbf{C}$. Thus, γ extends holomorphically to $\{\lambda \mid |\lambda| \leq 1\}$; let us choose such an extension and call it γ'. We construct an element $\hat{\gamma}$ of $\Lambda_{I,\mathbf{R}}GL_n\mathbf{C}$ by defining

$$\hat{\gamma}(\lambda) = \begin{cases} \gamma'(\lambda) & \text{for } |\lambda| = \epsilon \\ \gamma'(1/\bar{\lambda})^{-1*} & \text{for} |\lambda| = 1/\epsilon. \end{cases}$$

(4) Let $F : \mathbf{C} \to \Omega U_n$ be a solution of (Ω). As explained in Chapter 16, we may consider F as a map from $\mathbf{C}$ to $\Lambda^1_{E,\mathbf{R}}GL_n\mathbf{C}$.

We are now in a position to state the relationship between our two actions:

Theorem. *If W corresponds to F, then $\gamma \cdot W$ corresponds to $\hat{\gamma} \cdot F$.*

Proof. To find $\hat{\gamma} \cdot F$, we must find the "$E - I$" factorization of $\hat{\gamma}F(z,)$: $C^\epsilon \cup C^{1/\epsilon} \to GL_n\mathbf{C}$. Fortunately we are in the special situation considered at the end of Chapter 12. Namely, $\hat{\gamma}F(z,)$ is defined for all λ with $\epsilon \leq |\lambda| \leq 1$. Hence $\hat{\gamma} \cdot F$ is given by the first factor of

$$\hat{\gamma}F(z,)|_{|\lambda|=1}$$

with respect to the decomposition $\Lambda GL_n\mathbf{C} = \Omega U_n\Lambda_+GL_n\mathbf{C}$. This is simply $(\gamma F)_u$. By the remarks above, this corresponds to $\gamma \cdot W$. ∎

We have shown that the new action of $\Lambda_+GL_n\mathbf{C}$ corresponds to the old action of a *subgroup* of $\Lambda_{I,\mathbf{R}}GL_n\mathbf{C}$. Since the new action of ΩU_n and the old action of $\Lambda_{E,\mathbf{R}}GL_n\mathbf{C}$ are just given by translation, it follows that the new action of $\Lambda GL_n\mathbf{C}$ $(= \Omega U_n\Lambda_+GL_n\mathbf{C})$ is entirely determined by the old action of $\Lambda^{\epsilon,1/\epsilon}_{\mathbf{R}}GL_n\mathbf{C}$ $(= \Lambda^1_{E,\mathbf{R}}GL_n\mathbf{C}\,\Lambda_{I,\mathbf{R}}GL_n\mathbf{C})$. We do not claim that the converse is true, in general – we return to this matter in section III.

II. An example.

Let us return briefly to Example (2) of Chapter 16, i.e., the solution $F(z,\lambda) = \pi_{f(z)} + \lambda\pi_{f(z)}^\perp$ of (Ω). The corresponding solution W of (Gr) is given by

$$W(z) = F(z,\lambda)H_+^{(n)} = f(z) \oplus \lambda H_+^{(n)}.$$

The new action makes it very easy to obtain the results of section IV of Chapter 16. First we have:

Proposition. *For $X \in GL_n\mathbf{C}$, $X \cdot (\pi_f + \lambda\pi_f{}^\perp) = \pi_{Xf} + \lambda\pi_{Xf}{}^\perp$.*

Proof. $X(f(z) \oplus \lambda H_+^{(n)}) = Xf(z) \oplus \lambda H_+^{(n)}$. ∎

Next, we prove that the subgroup

$$\Lambda'_+ GL_n\mathbf{C} = \{\gamma \in \Lambda_+ GL_n\mathbf{C} \mid \gamma(\lambda) = I + \sum_{i \geq 1} A_i\lambda^i\}$$

fixes W:

Theorem. *For $\gamma \in \Lambda'_+ GL_n\mathbf{C}$, we have $\gamma \cdot (\pi_f + \lambda\pi_f{}^\perp) = \pi_f + \lambda\pi_f{}^\perp$.*

Proof. $(I + \sum_{i \geq 1} A_i\lambda^i)(f(z) \oplus \lambda H_+^{(n)}) = f(z) \oplus \lambda H_+^{(n)}$. ∎

As in the case of the old action, we conclude that the orbit $\pi_f + \lambda\pi_f{}^\perp$ under $\Lambda_+ GL_n\mathbf{C}$ is given by the orbit of f under $GL_n\mathbf{C} \cong \Lambda_+ GL_n\mathbf{C}/\Lambda'_+ GL_n\mathbf{C}$.

III. Description of harmonic maps $S^2 \to U_n$ in terms of unitons.

One of the main results of Uhlenbeck [1989] concerning harmonic maps $\phi : S^2 \to U_n$ is:

Theorem A (Uhlenbeck [1989]). *Let $\phi : S^2 \to U_n$ be harmonic. Then there exists a corresponding solution F of (Ω) and an element γ of ΩU_n such that*

(i) $\gamma(\lambda)F(z,\lambda) = \sum_{i=0}^r A_i(z)\lambda^i$, i.e., γF is a polynomial in λ;

(ii) γF may be factorized in the form

$$\gamma(\lambda)F(\ ,\lambda) = (\pi_{f_1} + \lambda\pi_{f_1}{}^\perp)\ldots(\pi_{f_r} + \lambda\pi_{f_r}{}^\perp)$$

where $f_i : S^2 \to \mathrm{Gr}_{k_i}(\mathbf{C}^n)$ for some k_i;

(iii) and finally, $r \leq n - 1$. ∎

The first factor $\pi_{f_1} + \lambda\pi_{f_1}{}^\perp$ in part (ii) is of the type considered in the previous section, i.e., f_1 is holomorphic. This is called a *uniton* in Uhlenbeck [1989]. The maps f_i for $i \geq 2$ are not necessarily holomorphic; they are holomorphic in a "twisted" sense, and Uhlenbeck calls F an r-*uniton*. This description is not quite so useful as it seems, because of the difficulty of dealing with the twisting. However, an interesting application of this description (Wood [1989]) is that any harmonic map $S^2 \to U_n$ can be constructed by performing elementary operations (algebraic operations and integrations) on meromorphic functions. We prove a version of this result in Chapter 22.

Part (i) of the above theorem is the fundamental "finiteness property" of harmonic maps defined on S^2. In Segal [1989], a different version of this was proved:

Theorem B (Segal [1989]). *Let* $\phi : S^2 \to U_n$ *be harmonic. Then there exists a corresponding solution* W *of* (Gr) *and an element* γ *of* $\Lambda GL_n\mathbf{C}$ *such that the image of* $\gamma W : S^2 \to \mathrm{Gr}^{(n)}$ *lies in the algebraic Grassmannian*

$$\mathrm{Gr}^{(n),\mathrm{alg}} = \{W \subseteq H^{(n)} \mid \lambda W \subseteq W \text{ and } \lambda^r H_+^{(n)} \subseteq W \subseteq \lambda^{-r} H_+^{(n)} \text{ for some } r\}.$$

Furthermore, this element γ *may be chosen so that* $\lambda^r H_+^{(n)} \subseteq W \subseteq H_+^{(n)}$, *for some* r. ∎

The algebraic Grassmannian is the subspace of $\mathrm{Gr}^{(n)}$ which corresponds to the *algebraic loop group*

$$\Omega^{\mathrm{alg}} U_n = \{\gamma \in \Omega U_n \mid \gamma(\lambda) = \sum_{i=-r}^{r} A_i \lambda^i \text{ for some } r\}.$$

Theorem B is equivalent to part (i) of Theorem A. Segal deduces parts (ii) and (iii) of Theorem A from this by considering the flag

$$W = W_{(r)} \subseteq W_{(r-1)} \subseteq \ldots \subseteq W_{(1)} \subseteq H_+^{(n)}$$

where $W_{(r-i)} = \lambda^{-i} W \cap H_+^{(n)}$.

The number r appearing in Theorems A and B is called the *uniton number* of ϕ. (More precisely, the least such number is called the *minimal uniton number* of ϕ.) In general, a harmonic map $\phi : \mathbf{C} \to U_n$ is said to have *finite uniton number* if there exists a corresponding solution F of (Ω) which (possibly after multiplication by an element of ΩU_n) is polynomial in λ. Theorems A and B say that any harmonic map which extends to S^2 has finite uniton number.

It is shown in §5 of Guest and Ohnita [1993] that, for harmonic maps of finite uniton number, the old action of $\Lambda_{\mathbf{R}}^{\epsilon,1/\epsilon} GL_n\mathbf{C}$ is determined by the new action of $\Lambda GL_n\mathbf{C}$. Thus, in this case, the old and new actions are completely equivalent.

IV. The algebraic Grassmannian $\mathrm{Gr}^{(n),\mathrm{alg}}$.

The subspace $\mathrm{Gr}^{(n),\mathrm{alg}}$ of $\mathrm{Gr}^{(n)}$ (defined in Theorem B above) is a "purely algebraic object", in contrast to $\mathrm{Gr}^{(n)}$ itself. This fact is responsible for the special properties of harmonic maps $S^2 \to U_n$. The space $\mathrm{Gr}^{(n),\mathrm{alg}}$ is a very important manifestation of the "hidden" algebraic structure of the harmonic map equation, and a typical illustration of algebraic methods in the theory of integrable systems. In Chapters 20-22 we try to justify these assertions in more detail, so we end this chapter with some further remarks on $\mathrm{Gr}^{(n),\mathrm{alg}}$.

We have mentioned two group actions on $\mathrm{Gr}^{(n)}$ so far, the natural action of $\Lambda GL_n\mathbf{C}$, and the action of S^1 given by "rotation of parameter". Together, these give an action of the semi-direct product $S^1 \ltimes \Lambda GL_n\mathbf{C}$, if we use the formula

$$(u \cdot \gamma) \cdot W = u \cdot (\gamma \cdot W).$$

This is a slightly larger group than the group $S^1 \ltimes \Lambda U_n$ which arose in our discussion of affine Lie algebras, in Chapter 11. However, it is smaller than the naive candidate for the complexification of $S^1 \ltimes \Lambda U_n$, namely "$\mathbf{C}^* \ltimes \Lambda GL_n\mathbf{C}$". Unfortunately, the latter is not well defined, because the "action" of $\mathbf{C}^*$ on $\Lambda GL_n\mathbf{C}$ given by $(u \cdot \gamma)(\lambda) = \gamma(u\lambda)$ may not be defined when $|u| \neq 1$.

A resolution of this difficulty is provided by the algebraic loop group

$$\Lambda^{\mathrm{alg}}GL_n\mathbf{C} = \{\gamma \in \Lambda GL_n\mathbf{C} \mid \gamma(\lambda) = \sum_{i=-r}^{r} A_i\lambda^i, \; \gamma(\lambda)^{-1} = \sum_{i=-s}^{s} B_i\lambda^i\}$$

(cf. the definition of $\Omega^{\mathrm{alg}}U_n$ in the last section). The group $\mathbf{C}^*$ acts on $\Lambda^{\mathrm{alg}}GL_n\mathbf{C}$ by the formula $(u \cdot \gamma)(\lambda) = \gamma(u\lambda)$, and we have a well defined semi-direct product $\mathbf{C}^* \ltimes \Lambda^{\mathrm{alg}}GL_n\mathbf{C}$.

The group $\Lambda^{\mathrm{alg}}GL_n\mathbf{C}$ acts on $\mathrm{Gr}^{(n),\mathrm{alg}}$. Not surprisingly, we have the following "algebraic" version of Theorem 1 of Chapter 12:

$$\mathrm{Gr}^{(n),\mathrm{alg}} \cong \Lambda^{\mathrm{alg}}U_n/U_n \cong \Lambda^{\mathrm{alg}}GL_n\mathbf{C}/\Lambda_+^{\mathrm{alg}}GL_n\mathbf{C}.$$

There is a similar analogue of Theorem 2 of Chapter 12:

$$\mathrm{Fl}^{(n),\mathrm{alg}} \cong \Lambda^{\mathrm{alg}}U_n/T_n \cong \Lambda^{\mathrm{alg}}GL_n\mathbf{C}/\Lambda_+^{\Delta,\mathrm{alg}}GL_n\mathbf{C}.$$

These are proved by the same method as for Theorems 1 and 2. In Chapter 19 we make use of the "full" complexified group $\mathbf{C}^* \ltimes \Lambda^{\mathrm{alg}}GL_n\mathbf{C}$. It is easy to verify that this group, like its real form, acts as a symmetry group of the harmonic map equation (Gr), in the case of harmonic maps $S^2 \to U_n$.

There is a beautiful Morse theoretic explanation of the difference between $\mathrm{Gr}^{(n)}$ and $\mathrm{Gr}^{(n),\mathrm{alg}}$. This is closely related to our discussion of "height functions on Ad-orbits" in Chapter 4, and to the discussion of Bruhat decompositions of Grassmannians in Chapter 14. It depends on the fact that ΩU_n, and hence $\mathrm{Gr}^{(n)}$, can be identified with the Ad-orbit $M_{(\sqrt{-1},0)}$ in $\sqrt{-1}\mathbf{R} \ltimes \Lambda u_n$ (see Chapter 11). The height function $H^{(\sqrt{-1},0)} : \Omega U_n \to \mathbf{R}$ turns out to be (up to constants) the energy function $\gamma \mapsto \int ||\gamma'||^2$. It is well known that the critical points of this functional are the homomorphisms $S^1 \to U_n$. The Hamiltonian vector field is generated by the action of the subgroup $S^1 \times \{I\}$ of $S^1 \ltimes \Lambda U_n$; the gradient vector field is generated by the action of the multiplicative subgroup $(0, \infty) \times \{I\}$ of $\mathbf{C}^* \ltimes \Lambda^{\mathrm{alg}} GL_n \mathbf{C}$.

The main difference between this (infinite dimensional) example and the (finite dimensional) examples of Chapter 4 is that the group $(0, \infty) \times \{I\}$ does not act on the whole of ΩU_n. In other words, the integral curves of the gradient vector field are not necessarily defined for all time. This is due to the non-compactness of ΩU_n. It turns out that the semigroup $[1, \infty) \times \{I\}$ – which represents the "downwards" gradient flow – acts on the whole of ΩU_n, but the semigroup $(0, 1] \times \{I\}$ – which represents the "upwards" gradient flow – acts only on the algebraic loop group $\Omega^{\mathrm{alg}} U_n$. This reflects the analytical fact that the energy functional is bounded below, but not above. More generally, the semigroup

$$\mathbf{C}^*_{\geq 1} = \{(u, I) \mid u \in \mathbf{C}^*, |u| \geq 1\}$$

acts on all of ΩU_n, but the semigroup

$$\mathbf{C}^*_{\leq 1} = \{(u, I) \mid u \in \mathbf{C}^*, |u| \leq 1\}$$

acts only on $\Omega^{\mathrm{alg}} U_n$.

In section III of Chapter 14 we considered the two (dual) Bruhat decompositions of $\mathrm{Gr}_i(\mathbf{C}^n)$, and their Morse theoretic interpretation. That theory extends to our infinite dimensional Grassmannian, if we take account of the difference between $\mathrm{Gr}^{(n)}$ and $\mathrm{Gr}^{(n),\mathrm{alg}}$. If $\gamma : S^1 \to U_n$ is a homomorphism (i.e., a critical point of the energy function), it can be shown that the stable manifold containing γ is the orbit

$$\Lambda_- GL_n \mathbf{C} \cdot \gamma H^{(n)}_+.$$

This is the same as

$$\Lambda_- GL_n \mathbf{C} \begin{pmatrix} \lambda^{k_1} & & \\ & \ddots & \\ & & \lambda^{k_n} \end{pmatrix} H^{(n)}_+$$

for unique integers $k_1 \geq \cdots \geq k_n$. We therefore have a (disjoint) decomposition

$$\mathrm{Gr}^{(n)} = \bigcup_{k_1 \geq \cdots \geq k_n} \Lambda_- GL_n \mathbf{C} \begin{pmatrix} \lambda^{k_1} & & \\ & \ddots & \\ & & \lambda^{k_n} \end{pmatrix} H^{(n)}_+.$$

It is instructive to consider the action of $\mathbf{C}^*_{\geq 1}$ from this point of view. We have

$$u \cdot \sum_{i \leq 0} A_i \lambda^i \begin{pmatrix} \lambda^{k_1} & & \\ & \ddots & \\ & & \lambda^{k_n} \end{pmatrix} H^{(n)}_+ = \sum_{i \leq 0} A_i (u\lambda)^i \begin{pmatrix} \lambda^{k_1} & & \\ & \ddots & \\ & & \lambda^{k_n} \end{pmatrix} H^{(n)}_+.$$

As $|u| \to \infty$, we have

$$u \cdot \sum_{i \leq 0} A_i \lambda^i \begin{pmatrix} \lambda^{k_1} & & \\ & \ddots & \\ & & \lambda^{k_n} \end{pmatrix} H_+^{(n)} \longrightarrow A_0 \begin{pmatrix} \lambda^{k_1} & & \\ & \ddots & \\ & & \lambda^{k_n} \end{pmatrix} H_+^{(n)},$$

so we see explicitly how any point flows to a homomorphism.

There is a corresponding decomposition of $\Lambda GL_n \mathbf{C}$:

$$\Lambda GL_n \mathbf{C} = \bigcup_{k_1 \geq \cdots \geq k_n} \Lambda_- GL_n \mathbf{C} \begin{pmatrix} \lambda^{k_1} & & \\ & \ddots & \\ & & \lambda^{k_n} \end{pmatrix} \Lambda_+ GL_n \mathbf{C}.$$

These decompositions of $\mathrm{Gr}^{(n)}$ and $\Lambda GL_n \mathbf{C}$ are referred to as *Birkhoff decompositions*. (This is a refinement of our terminology in Chapter 12, where we called the "partial decomposition"

$$\Lambda GL_n \mathbf{C} \supseteq \Lambda_-^1 GL_n \mathbf{C} \, \Lambda_+ GL_n \mathbf{C}$$

the Birkhoff decomposition. The factorization $\gamma = \gamma_- \gamma_0 \gamma_+$ of Chapter 12 arises from this "new" Birkhoff decomposition.)

The unstable manifold containing γ turns out to be the orbit of γ under $\Lambda_+ GL_n \mathbf{C}$, but there are two new features now:

(1) $\Lambda GL_n \mathbf{C}$ is strictly larger than $\bigcup_\gamma \Lambda_+ GL_n \mathbf{C} \cdot \gamma$

(2) $\Lambda_+ GL_n \mathbf{C} \cdot \gamma = \Lambda_+^{\mathrm{alg}} GL_n \mathbf{C} \cdot \gamma$ (and this is a finite dimensional manifold).

Thus, instead of a decomposition of $\mathrm{Gr}^{(n)}$, the unstable manifolds give a decomposition of $\mathrm{Gr}^{(n),\mathrm{alg}}$:

$$\mathrm{Gr}^{(n),\mathrm{alg}} = \bigcup_{k_1 \geq \cdots \geq k_n} \Lambda_+^{\mathrm{alg}} GL_n \mathbf{C} \begin{pmatrix} \lambda^{k_1} & & \\ & \ddots & \\ & & \lambda^{k_n} \end{pmatrix} H_+^{(n)}.$$

The action of $\mathbf{C}_{\leq 1}^*$ on a piece of this decomposition is given by

$$u \cdot \sum_{i \geq 0} A_i \lambda^i \begin{pmatrix} \lambda^{k_1} & & \\ & \ddots & \\ & & \lambda^{k_n} \end{pmatrix} H_+^{(n)} = \sum_{i \geq 0} A_i (u\lambda)^i \begin{pmatrix} \lambda^{k_1} & & \\ & \ddots & \\ & & \lambda^{k_n} \end{pmatrix} H_+^{(n)},$$

and, as $|u| \to 0$, we have

$$u \cdot \sum_{i \geq 0} A_i \lambda^i \begin{pmatrix} \lambda^{k_1} & & \\ & \ddots & \\ & & \lambda^{k_n} \end{pmatrix} H_+^{(n)} \longrightarrow A_0 \begin{pmatrix} \lambda^{k_1} & & \\ & \ddots & \\ & & \lambda^{k_n} \end{pmatrix} H_+^{(n)}.$$

The corresponding group decomposition is

$$
\Lambda^{\mathrm{alg}} GL_n \mathbf{C} = \bigcup_{k_1 \geq \cdots \geq k_n} \Lambda^{\mathrm{alg}}_+ GL_n \mathbf{C} \begin{pmatrix} \lambda^{k_1} & & \\ & \ddots & \\ & & \lambda^{k_n} \end{pmatrix} \Lambda^{\mathrm{alg}}_+ GL_n \mathbf{C}.
$$

These decompositions are referred to as *Bruhat decompositions*.

Remark: The orbit $\Lambda^{\mathrm{alg}}_+ GL_n \mathbf{C} \cdot \gamma$ is not a cell, because γ is not (usually) an isolated critical point. In fact, the connected critical manifold containing γ is the orbit $GL_n \mathbf{C} \cdot \gamma$, i.e., the conjugacy class of γ in ΩU_n, and $\Lambda^{\mathrm{alg}}_+ GL_n \mathbf{C} \cdot \gamma$ has the structure of a vector bundle over $GL_n \mathbf{C} \cdot \gamma$. The finite dimensional analogue of this (in Chapter 14) would be obtained by replacing N_n (or Δ_n) by a larger parabolic subgroup of G^c. Similar remarks apply to $\Lambda_- GL_n \mathbf{C} \cdot \gamma$.

Bibliographical comments for Chapters 16-17.

Chapter 16 presents the approach to harmonic maps which was introduced in Uhlenbeck [1989]. We have emphasized the loop group theoretic point of view, as in Bergvelt and Guest [1991].

Some explicit computations can be done by making use of loops of "simplest type" (Uhlenbeck [1989]; Bergvelt and Guest [1991]). A related explicit construction, the Darboux transformation, has been studied in Gu and Hu [1995].

The action of S^1 was first noted by Terng (see §7 of Uhlenbeck [1989]). In the example of section IV of Chapter 16, it is easy to see that this action is trivial. However, it is certainly not trivial on more general harmonic maps (see Chapter 20).

The Grassmannian theoretic point of view was introduced in Segal [1989]. The action of $\Lambda GL_n \mathbf{C}$ on solutions of (Gr), and its relationship with the dressing action of Uhlenbeck [1989], was studied in Guest and Ohnita [1993].

The various formulations of the harmonic map equation – (Ω) and (Λ) in Chapter 16, (Gr) in Chapter 17 – are extremely useful. Each expresses the harmonic map equation in terms of an infinite dimensional manifold, but each has its own special flavour. We have already pointed out that (Λ) is particularly suitable when discussing the S^1-action, for example. On the other hand, the geometrical nature of (Gr) is very convenient for explicit calculations. For example, the treatment of the example in section II of Chapter 17 was much easier than in section IV of Chapter 16. We see further examples in Chapter 20.

A comment should be made here on the (omitted) proofs of Theorems A and B, the finiteness theorems for harmonic maps defined on S^2. Uhlenbeck proves Theorem A by observing that, for any solution $F : S^2 \to \Omega U_n$ of (Ω), the coefficients of F satisfy a certain elliptic partial differential equation. The compactness of S^2 leads to the desired result. Segal proves Theorem B by showing more generally that, for any *holomorphic* map $W : S^2 \to \mathrm{Gr}^{(n)}$, there exists an element γ of of $\Lambda GL_n \mathbb{C}$ such that the image of γW lies in $\mathrm{Gr}^{(n),\mathrm{alg}}$. Again, compactness of S^2 is an essential ingredient. Both proofs apply also to the case of a map $F : M \to \Omega U_n$ (or $W : M \to \mathrm{Gr}^{(n)}$), for any compact Riemann surface M. However, this says nothing about harmonic maps $\phi : M \to U_n$, as there is no guarantee that ϕ corresponds to such an F. The construction of Chapter 16 would merely produce a map $F : \tilde{M} \to \Omega U_n$, where $\tilde{M}$ is the universal cover of M – but $\tilde{M}$ is compact only if $M = S^2$.

The relationship between $\mathrm{Gr}^{(n)}$ and $\mathrm{Gr}^{(n),\mathrm{alg}}$, and all the related Morse theory results, are explained in detail in Pressley [1982] and Chapter 8 of Pressley and Segal [1986]. The first Morse theoretic approach to ΩG was that of Bott [1956].

Chapter 18: Harmonic maps from C to a symmetric space

A *symmetric space* is a homogeneous space G/K such that

(1) there is an involution $\sigma : G \to G$ (i.e., an automorphism $\sigma : G \to G$ of order 2), such that

(2) $(G_\sigma)_0 \subseteq K \subseteq G_\sigma$, where $(G_\sigma)_0$ is the identity component of $G_\sigma = \{g \in G \mid \sigma(g) = g\}$.

If σ is an inner involution (i.e., $\sigma(g) = aga^{-1}$ for some $a \in G$), then we say that G/K is an *inner symmetric space*. We assume that G is a compact connected Lie group here; hence G/K is also compact and connected.

It is well known that the map

$$i : G/K \to G, \quad gK \mapsto \sigma(g)g^{-1}$$

defines an immersion of G/K into G (this is called the *Cartan immersion*). If we define

$$N^\sigma = \{x \in G \mid \sigma(x) = x^{-1}\}$$

then it is known that $i(G/K)(\cong G/G_\sigma)$ is the component of N^σ which contains the identity element e of G.

Examples:

(1) Let $G = U_n$, $K = U_k \times U_{n-k}$. Let

$$\sigma(X) = E_k X E_k^{-1}, \quad \text{where } E_k = \begin{pmatrix} I_k & 0 \\ 0 & -I_{n-k} \end{pmatrix}$$

and where I_k denotes the $k \times k$ identity matrix. The symmetric space G/K is the Grassmannian $\mathrm{Gr}_k(\mathbf{C}^n)$. In this example we have $(G_\sigma)_0 = K = G_\sigma$. We also have

$$\begin{aligned} N^\sigma &= \{X \in U_n \mid E_k X E_k^{-1} = X^{-1}\} \\ &= \{X \in U_n \mid (E_k X)^2 = I\}. \end{aligned}$$

Since the eigenvalues of $E_k X$ are 1 or -1 (when $X \in N^\sigma$), it is easy to see that N^σ has $n+1$ connected components; the components are indexed by the number of positive eigenvalues of $E_k X$. The k-th component may be identified with $\mathrm{Gr}_k(\mathbf{C}^n)$. The embedding of this component in U_n is given explicitly by

$$\mathrm{Gr}_k(\mathbf{C}^n) \to U_n, \quad V \mapsto E_k(\pi_V - \pi_{V^\perp})$$

where $\pi_V : \mathbf{C}^n \to \mathbf{C}^n$ denotes the Hermitian projection operator onto the k-plane V.

(2) Let $G = SO_{n+1}$, and let

$$\sigma(X) = E_n X E_n^{-1}, \quad \text{where } E_n = \begin{pmatrix} I_n & 0 \\ 0 & -1 \end{pmatrix}.$$

We have $G_\sigma = S(O_n \times O_1)$, which has two connected components. If we take $K = G_\sigma$, we obtain the symmetric space $G/K = \mathbf{R}P^n$, which is a "real" version of Example (1). But if we take $K = (G_\sigma)_0$, we obtain the symmetric space $G/K = S^n$.

It is well known that each component of N^σ is a totally geodesic submanifold of G. From this (see the comments in Example (2) of Chapter 16) one can derive the following basic fact concerning harmonic maps:

Proposition. $\phi : \mathbf{C} \to G/K$ *is harmonic if and only if* $i \circ \phi : \mathbf{C} \to G$ *is harmonic.* ∎

The proposition shows that the theory of harmonic maps $\mathbf{C} \to G/K$ is a special case of the theory of harmonic maps $\mathbf{C} \to G$. In this section we shall study this special case, using loop groups.

Not surprisingly, we shall use the twisted loop group

$$(\Lambda G)_\sigma = \{\gamma \in \Lambda G \mid \sigma(\gamma(\lambda)) = \gamma(-\lambda)\} \subseteq \Lambda G$$

where σ is the above involution. In Chapter 16, we saw that the harmonic map equation (for $\phi : \mathbf{C} \to G$) was equivalent to a "simpler" system of equations (for $F : \mathbf{C} \to \Lambda G$), namely (Λ). Let us now consider the twisted version of this system, i.e.,

$$(\Lambda_\sigma) \qquad \begin{aligned} F^{-1}F_z &= \text{linear in } \tfrac{1}{\lambda} \ (= U + \tfrac{1}{\lambda}V \text{ for some } U, V) \\ F^{-1}F_{\bar{z}} &= \text{linear in } \lambda \ (= X + \lambda Y \text{ for some } X, Y) \end{aligned}$$

where $F : \mathbf{C} \to (\Lambda G)_\sigma$. (Since $F^{-1}F_z$, $F^{-1}F_{\bar{z}}$ take values in $(\Lambda \mathbf{g})_\sigma \otimes \mathbf{C}$, it follows that U, X take values in the $(+1)$-eigenspace of $\mathbf{g} \otimes \mathbf{C}$, and that V, Y take values in the (-1)-eigenspace.) It turns out that this system is equivalent to the harmonic map equation for maps $\mathbf{C} \to G/K$:

Proposition.

(1) Let $F : \mathbf{C} \to (\Lambda G)_\sigma$ be a solution of (Λ_σ). Then $\phi(z) = [F(z, 1)]$ defines a harmonic map from $\mathbf{C}$ to G/K.

(2) Conversely, if ϕ is a harmonic map from $\mathbf{C}$ to G/K, then there exists a solution F of (Λ_σ) such that $\phi(z) = [F(z, 1)]$.

Proof. Any (smooth) map $\phi : \mathbf{C} \to G/K$ may be written in the form $\phi = [\psi]$, for some map $\psi : \mathbf{C} \to G$. (This is because $G \to G/K$ is a locally trivial fibre bundle, and **C** is simply connected.) From the above discussion, ϕ is harmonic if and only if $\theta = \sigma(\psi)\psi^{-1}$ is harmonic.

Let M', N' be the components of $\psi^{-1}\psi_z$ in the $(-1), (+1)$-eigenspaces of σ (on $\mathbf{g} \otimes \mathbf{C}$), and let M'', N'' be the analogous components of $\psi^{-1}\psi_{\bar{z}}$. We have

$$\theta^{-1}\theta_z = \psi\sigma(\psi^{-1})(\sigma(\psi_z)\psi^{-1} - \sigma(\psi)\psi^{-1}\psi_z\psi^{-1})$$
$$= \psi(\sigma(\psi^{-1}\psi_z) - \psi^{-1}\psi_z)\psi^{-1}$$
$$= -2\psi M'\psi^{-1}$$

and

$$(\theta^{-1}\theta_z)_{\bar{z}} = -2\{\psi_{\bar{z}}M'\psi^{-1} + \psi M'_{\bar{z}}\psi^{-1} - \psi M'\psi^{-1}\psi_{\bar{z}}\psi^{-1}\}$$
$$= -2\psi\{(M'' + N'')M' + M'_{\bar{z}} - M'(M'' + N'')\}\psi^{-1}$$
$$= -2\psi\{[M'', M'] + [N'', M'] + M'_{\bar{z}}\}\psi^{-1}.$$

Similarly we have $(\theta^{-1}\theta_{\bar{z}})_z = -2\psi\{[M', M''] + [N', M''] + M''_z\}\psi^{-1}$, so the harmonic map equation for θ becomes

$$M'_{\bar{z}} + M''_z = [M', N''] + [M'', N'].$$

Using the same notation, the components of the identity $(\theta^{-1}\theta_z)_{\bar{z}} - (\theta^{-1}\theta_{\bar{z}})_z = [\theta^{-1}\theta_z, \theta^{-1}\theta_{\bar{z}}]$ in the eigenspaces of σ become

$$M'_{\bar{z}} - M''_z = [M', N''] - [M'', N']$$
$$N'_{\bar{z}} - N''_z = [M', M''] + [N', N''].$$

It is easily verified that the last three equations are equivalent to the "zero-curvature equation with parameter" $(A_\lambda)_{\bar{z}} - (B_\lambda)_z = [A_\lambda, B_\lambda]$, where

$$A_\lambda = \tfrac{1}{\lambda}M' + N', \quad B_\lambda = \lambda M'' + N''.$$

This equation is equivalent to (Λ_σ), by the usual argument (see Chapter 9). ∎

Thus, the theory of harmonic maps $\mathbf{C} \to G/K$ (or $\mathbf{C} \to G/G_\sigma$) can be developed just like the theory of harmonic maps $\mathbf{C} \to G$, if we use the twisted loop group $(\Lambda G)_\sigma$ instead of the loop group ΛG.

We can reformulate the harmonic map equation in terms of the "twisted based loop group"

$$(\Omega G)_\sigma = \{\gamma \in \Omega G \mid \sigma(\gamma(\lambda)) = \gamma(-\lambda)\gamma(-1)^{-1}\}.$$

Consider the following system

(Ω_σ)
$$F^{-1}F_z = \text{linear in } \tfrac{1}{\lambda}$$
$$F^{-1}F_{\bar{z}} = \text{linear in } \lambda$$

where $F : \mathbf{C} \to (\Omega G)_\sigma$.

Corollary.

(1) Let $F : \mathbf{C} \to (\Omega G)_\sigma$ be a solution of (Ω_σ). Then $\phi(z) = F(z, -1)$ defines a harmonic map from $\mathbf{C}$ to (a component of) N^σ.

(2) Conversely, if ϕ is a harmonic map from $\mathbf{C}$ to (a component of) N^σ, then there exists a solution F of (Ω_σ) such that $\phi(z) = F(z, -1)$.

Proof. (1) Any solution F of (Ω_σ) is automatically a solution of (Ω), so the map $F(\ , -1) = \phi : \mathbf{C} \to G$ is a harmonic map. We have to show that $\phi(\mathbf{C}) \subseteq N^\sigma$. Since $F(z, \ ) \in (\Omega G)_\sigma$, we have

$$\sigma\phi(z) = \sigma F(z, -1) = F(z, 1)F(z, -1)^{-1} = F(z, -1)^{-1} = \phi(z)^{-1},$$

as required.

(2) Let ϕ be a harmonic map from $\mathbf{C}$ to N^σ, and let $\psi : \mathbf{C} \to G$ be such that $\phi = \sigma(\psi)\psi^{-1}$. By the (proof of the) proposition, there exists a solution F of (Λ_σ) such that $F(z, 1) = \psi$. It follows that $F(z, -1) = \sigma(\psi)$. The map $E = F\psi^{-1}$ is a solution of (Ω_σ) such that $E(z, -1) = \phi(z)$. $\blacksquare$

Examples:

(3) Consider G, K, σ as in Example (1) above, so that $G/K = \mathrm{Gr}_k(\mathbf{C}^n)$. Let $f : \mathbf{C} \to \mathrm{Gr}_k(\mathbf{C}^n)$ be a holomorphic map. As we remarked in Chapter 16, it is well known that f is harmonic. Let us now try to find corresponding solutions of (Λ_σ) and (Ω_σ), as predicted by the proposition and its corollary.

We can write

$$f(z) = \psi(z)\,\mathrm{Span}\{e_1, \ldots, e_k\} \in \mathrm{Gr}_k(\mathbf{C}^n)$$

for some map $\psi : \mathbf{C} \to U_n$. Composition with the Cartan embedding gives the map

$$\mathbf{C} \to U_n, \quad z \mapsto E_k \psi(z) E_k^{-1} \psi(z)^{-1}.$$

We have $E_k = P_k - P_k^\perp$, where $P_k, P_k^\perp$ denote the Hermitian projection operators onto the subspaces $\mathrm{Span}\{e_1, \ldots, e_k\}$, $\mathrm{Span}\{e_{k+1}, \ldots, e_n\}$, respectively. Hence

$$\psi(z)E_k^{-1}\psi(z)^{-1} = \psi(z)(P_k - P_k^\perp)\psi(z)^{-1} = \pi_{f(z)} - \pi_{f(z)^\perp}$$

and our map is

$$\mathbf{C} \to U_n, \quad z \mapsto (P_k - P_k^\perp)(\pi_{f(z)} - \pi_{f(z)^\perp}).$$

Modulo the unimportant constant factor $P_k - P_k^\perp$, this is the harmonic map which we considered in Example (2) of Chapter 16. There, we show that

$F(\ ,\lambda) = \pi_f + \lambda\pi_{f\perp}$ is a solution of (Ω) corresponding to the harmonic map $\pi_f - \pi_{f\perp}$.

We claim that

$$F_1(\ ,\lambda) = (P_k + \tfrac{1}{\lambda}P_k^\perp)(\pi_f + \lambda\pi_{f\perp})$$

is a solution of (Ω_σ) which corresponds to our harmonic map. (Proof: Since $F = \pi_f + \lambda\pi_{f\perp}$ is a solution of (Ω), and $P_k + \tfrac{1}{\lambda}P_k^\perp$ is a constant loop, it suffices to show that F_1 takes values in $(\Omega U_n)_\sigma$. This may be verified by a routine calculation.)

A further modification produces a corresponding solution of (Λ_σ). We claim that

$$F_2(\ ,\lambda) = (P_k + \tfrac{1}{\lambda}P_k^\perp)\psi(P_k + \lambda P_k^\perp)$$

is the required map. (Proof: Observe that $F_2 = F_1\psi$. Since F_1 is a solution of (Ω_σ), and hence also of (Ω) and (Λ), it follows that F_2 is a solution of (Λ). So it remains to show now that F_2 takes values in $(\Lambda U_n)_\sigma$, i.e., that $\sigma F_2(z,\lambda) = F_2(z,-\lambda)$. This is obvious.) Note that the formula $F_2 = F_1\psi$ is consistent with the proof of the corollary.

(4) Consider the symmetric space $G \times G/\Delta$, where $\Delta = \{(g,g) \mid g \in G\}$, with the involution $\sigma(g,h) = (h,g)$. We have $(G \times G)_\sigma = \Delta$ and $N^\sigma = \{(g,g^{-1}) \mid g \in G\} \cong G$, both of which are connected. We claim that the system (Λ_σ) for $(G \times G)/\Delta$ is equivalent to the system (Λ) for G – which shows that the theory of harmonic maps $\mathbf{C} \to G$ can be considered as a special case of the theory of harmonic maps $\mathbf{C} \to G/K$! To prove this, we consider a solution $(E,F) : \mathbf{C} \to (\Lambda(G \times G))_\sigma$ of (Λ_σ). By definition we have $F(\lambda) = E(-\lambda)$, and

$$(E,F)^{-1}(E,F)_z = (A,A) + \tfrac{1}{\lambda}(B,-B),$$

so F is a solution of (Λ) for G. Conversely, if F is a solution of (Λ), then we obtain a solution (E,F) of (Λ_σ) by taking $E(\lambda) = F(-\lambda)$.

In fact, a more precise result is possible, which explains the special form of the harmonic map equation (Ω). Any (smooth) map $\mathbf{C} \to G \times G/\Delta$ may be represented by $\psi = (\phi,e) : \mathbf{C} \to G \times G$, for some $\phi : \mathbf{C} \to G$. For this particular ψ,

$$\psi^{-1}\psi_z = (\phi^{-1}\phi_z, 0)$$
$$= \tfrac{1}{2}(\phi^{-1}\phi_z, \phi^{-1}\phi_z) + \tfrac{1}{2}(\phi^{-1}\phi_z, -\phi^{-1}\phi_z)$$
$$= N' + M'.$$

For the corresponding solution (E,F) of (Λ_σ) we therefore have $F^{-1}F_z = \tfrac{1}{2}(1 - \lambda^{-1})\phi^{-1}\phi_z$. Thus, F is a solution of (Ω) for G.

Bibliographical comments.

See the comments at the end of Chapter 20.

Chapter 19: Harmonic maps from
C to a symmetric space (continued)

In this chapter we study the symmetry group of the harmonic map equation for maps into a symmetric space, in parallel with our earlier discussion for maps into a Lie group. In the next chapter we give two non-trivial applications of this theory.

I. The symmetry group for the harmonic map equation.

The twisted loop group

$$(\Lambda_{\mathbf{R}}^{\epsilon,1/\epsilon} G^c)_\sigma$$

acts as a symmetry group of the harmonic map equation. We use the decomposition

$$(\Lambda_{\mathbf{R}}^{\epsilon,1/\epsilon} G^c)_\sigma = (\Lambda_{E,\mathbf{R}} G^c)_\sigma (\Lambda_{I,\mathbf{R}}^{\hat{N}_\sigma,\hat{N}'_\sigma} G^c)_\sigma$$

where $\hat{N}_\sigma$ is the second factor in the Iwasawa decomposition $G_\sigma^c = G_\sigma \hat{N}_\sigma$ of G_σ^c. If $(\gamma^\epsilon, \gamma^{1/\epsilon}) \in (\Lambda_{\mathbf{R}}^{\epsilon,1/\epsilon} G^c)_\sigma$, then we write

$$(\gamma^\epsilon, \gamma^{1/\epsilon}) = (\gamma^\epsilon, \gamma^{1/\epsilon})_E (\gamma^\epsilon, \gamma^{1/\epsilon})_I$$

for the factorization with respect to this decomposition.

Let F be a solution of (Λ_σ). As usual we may assume that

$$F : \mathbf{C} \to (\Lambda_{E,\mathbf{R}} G^c)_\sigma,$$

i.e., that F is defined for all $\lambda \in E$. Let $(\gamma^\epsilon, \gamma^{1/\epsilon}) \in (\Lambda_{\mathbf{R}}^{\epsilon,1/\epsilon} G^c)_\sigma$. Then we define

$$(\gamma^\epsilon, \gamma^{1/\epsilon}) \cdot F = ((\gamma^\epsilon, \gamma^{1/\epsilon})F)_E.$$

As in the case of harmonic maps into a Lie group, we have:

Proposition. $(\gamma^\epsilon, \gamma^{1/\epsilon}) \cdot F$ *is also a solution of* (Λ_σ). ∎

We can construct families of harmonic maps by applying the action of this symmetry group to "obvious" harmonic maps, such as the examples given in Chapter 18. This time, however, we proceed directly with the "Grassmannian interpretation" of the harmonic map equation and the symmetry group.

II. The symmetry group from the Grassmannian point of view.

As in Chapter 17, we restrict ourselves to the case $G = U_n$. We take the involution σ to be

$$\sigma : U_n \to U_n, \quad \sigma(X) = E_k X E_k^{-1}$$

where E_k is a diagonal matrix with k entries equal to 1 and $n - k$ entries equal to -1. We may as well assume that

$$E_k = \begin{pmatrix} I_k & 0 \\ 0 & -I_{n-k} \end{pmatrix}$$

as in Example (1) of Chapter 18.

We have isomorphisms

$$T : \Lambda GL_n\mathbf{C} \to (\Lambda GL_n\mathbf{C})_\sigma, \quad T : \Lambda \mathbf{gl}_n\mathbf{C} \to (\Lambda \mathbf{gl}_n\mathbf{C})_\sigma$$

given by

$$T(\gamma)(\lambda) = \begin{pmatrix} I_k & 0 \\ 0 & \lambda I_{n-k} \end{pmatrix} \gamma(\lambda^2) \begin{pmatrix} I_k & 0 \\ 0 & \lambda I_{n-k} \end{pmatrix}^{-1}$$

(this is a slight generalization of the example of section II, Chapter 11). A crucial property of T is:

Lemma. *If F is a solution of (Λ_σ), then $T^{-1}(F)$ is a solution of (Λ).*

Proof. Let $A + \lambda^{-1}B, C + \lambda D \in (\Lambda \mathbf{gl}_n\mathbf{C})_\sigma$. We must show that $T^{-1}(A + \lambda^{-1}B)$, $T^{-1}(C + \lambda D)$ are linear in λ^{-1}, λ, respectively.

By definition of $(\Lambda \mathbf{gl}_n\mathbf{C})_\sigma$ we have

$$A + \tfrac{1}{\lambda}B = \begin{pmatrix} A_1 & 0 \\ 0 & A_2 \end{pmatrix} + \tfrac{1}{\lambda}\begin{pmatrix} 0 & B_1 \\ B_2 & 0 \end{pmatrix}.$$

It is easy to verify that

$$T^{-1}(A + \tfrac{1}{\lambda}B) = \begin{pmatrix} A_1 & 0 \\ 0 & A_2 \end{pmatrix} + \begin{pmatrix} 0 & B_1 \\ \tfrac{1}{\lambda}B_2 & 0 \end{pmatrix},$$

which is certainly linear in λ^{-1}.

Similarly, if

$$C + \lambda D = \begin{pmatrix} C_1 & 0 \\ 0 & C_2 \end{pmatrix} + \lambda \begin{pmatrix} 0 & D_1 \\ D_2 & 0 \end{pmatrix},$$

then we have

$$T^{-1}(C + \lambda D) = \begin{pmatrix} C_1 & 0 \\ 0 & C_2 \end{pmatrix} + \begin{pmatrix} 0 & \lambda D_1 \\ D_2 & 0 \end{pmatrix}$$

which is linear in λ. ∎

We wish to describe the subspace

$$(\Lambda GL_n \mathbf{C})_\sigma / (\Lambda_+ GL_n \mathbf{C})_\sigma \subseteq \Lambda GL_n \mathbf{C} / \Lambda_+ GL_n \mathbf{C}$$

in terms of the "Grassmannian model" $\mathrm{Gr}^{(n)} \cong \Lambda GL_n \mathbf{C} / \Lambda_+ GL_n \mathbf{C}$. The lemma encourages us to make use of T, so we observe next that T restricts to an isomorphism

$$T : \Lambda_+^P GL_n \mathbf{C} \to (\Lambda_+ GL_n \mathbf{C})_\sigma$$

where

$$\Lambda_+^P GL_n \mathbf{C} = \{\gamma \in \Lambda_+ GL_n \mathbf{C} \mid \gamma(0) \in P\}$$

and

$$P = \{X \in GL_n \mathbf{C} \mid X(\mathbf{C}^k)^\perp \subseteq (\mathbf{C}^k)^\perp \}.$$

Here, $\mathbf{C}^k$ denotes $\mathrm{Span}\{e_1, \ldots, e_k\}$. Thus, T induces an isomorphism

$$T : \Lambda GL_n \mathbf{C} / \Lambda_+^P GL_n \mathbf{C} \to (\Lambda GL_n \mathbf{C})_\sigma / (\Lambda_+ GL_n \mathbf{C})_\sigma.$$

We now define our Grassmannian model for $(\Lambda GL_n \mathbf{C})_\sigma / (\Lambda_+ GL_n \mathbf{C})_\sigma$:

Definition.

$$\mathrm{Fl}_{n-k,k}^{(n)} = \left\{ W_2 \subseteq W_1 \subseteq W_0 \,\middle|\, \begin{array}{l} W_0 \in \mathrm{Gr}^{(n)}, \ \lambda W_0 = W_2 \\ \dim W_1/W_2 = n - k, \dim W_0/W_1 = k \end{array} \right\}.$$

The group $\Lambda GL_n \mathbf{C}$ acts transitively on $\mathrm{Fl}_{n-k,k}^{(n)}$, and $\Lambda_+^P GL_n \mathbf{C}$ is the isotropy subgroup at the point

$$\lambda H_+^{(n)} \subseteq (\mathbf{C}^k)^\perp \oplus \lambda H_+^{(n)} \subseteq H_+^{(n)}.$$

This may be proved in the same way as (or deduced from) the corresponding fact for the flag manifold $\mathrm{Fl}^{(n)}$ in Chapter 12. We have

$$\mathrm{Fl}_{n-k,k}^{(n)} \cong \Lambda U_n / (U_{n-k} \times U_k) \cong \Lambda GL_n \mathbf{C} / \Lambda_+^P GL_n \mathbf{C}.$$

The relationship between $\mathrm{Fl}_{n-k,k}^{(n)}$ and $\mathrm{Fl}^{(n)}$ is analogous to the relationship between the Grassmannian $\mathrm{Gr}_{n-k}(\mathbf{C}^n)$ and the flag manifold $F_{1,2,\ldots,n-1}(\mathbf{C}^n)$ (cf. Chapter 3).

We obtain our desired description of $(\Lambda GL_n \mathbf{C})_\sigma / (\Lambda_+ GL_n \mathbf{C})_\sigma$:

Proposition. $(\Lambda GL_n\mathbf{C})_\sigma/(\Lambda_+GL_n\mathbf{C})_\sigma \cong \mathrm{Fl}^{(n)}_{n-k,k}.$ ∎

Just as the Grassmannian $\mathrm{Gr}^{(n)}$ is responsible for the Iwasawa decomposition of $\Lambda GL_n\mathbf{C}$ (see Chapter 12), the flag manifold $\mathrm{Fl}^{(n)}_{n-k,k}$ is responsible for the Iwasawa decomposition of $(\Lambda GL_n\mathbf{C})_\sigma$.

Remark: We have an embedding of $\mathrm{Fl}^{(n)}_{n-k,k}$ in $\mathrm{Gr}^{(n)}$, corresponding to the embedding of $(\Lambda GL_n\mathbf{C})_\sigma/(\Lambda_+GL_n\mathbf{C})_\sigma$ in $\Lambda GL_n\mathbf{C}/\Lambda_+GL_n\mathbf{C}$. This is most conveniently described via an identification $H^{(n)} \to H^{(1)}$, $\lambda^k e_i \mapsto \lambda^{nk+i-1}$. However, an explicit description of this embedding will not be needed here.

A solution F of (Λ_σ) defines a map $W_2 \subseteq W_1 \subseteq W_0 : \mathbf{C} \to \mathrm{Fl}^{(n)}_{n-k,k}$, where

$$W_2 = \lambda T^{-1}FH^{(n)}_+, \quad W_1 = T^{-1}F((\mathbf{C}^k)^\perp \oplus \lambda H^{(n)}_+), \quad W_0 = T^{-1}FH^{(n)}_+.$$

We shall abbreviate this map by $\{W_i\} : \mathbf{C} \to \mathrm{Fl}^{(n)}_{n-k,k}$. Conversely, any map $\{W_i\} : \mathbf{C} \to \mathrm{Fl}^{(n)}_{n-k,k}$ corresponds to some map $F : \mathbf{C} \to (\Lambda GL_n\mathbf{C})_\sigma$. We show next that the condition for $\{W_i\}$ to correspond to a solution of (Λ_σ) is:

(Fl)
$$(W_i)_z \subseteq W_{i-1}, \ i = 1, 2$$
$$(W_i)_{\bar{z}} \subseteq W_i, \ i = 0, 1, 2.$$

Proposition. *F is a solution of (Λ_σ) if and only if $\{W_i\}$ is a solution of* (Fl).

Proof. If F is a solution of (Λ_σ), then $T^{-1}F$ is a solution of (Λ), by the lemma, and so we have $(W_2)_z \subseteq W_0$. We need a little more than this, so let us write

$$(T^{-1}F)^{-1}(T^{-1}F)_z = \begin{pmatrix} A_1 & 0 \\ 0 & A_2 \end{pmatrix} + \begin{pmatrix} 0 & B_1 \\ \lambda^{-1}B_2 & 0 \end{pmatrix}$$

as in the proof of the lemma, and then compute $(W_2)_z$ more carefully:

$$\begin{aligned}
(W_2)_z &= \lambda(W_0)_z \\
&= \lambda(T^{-1}F)_z H^{(n)}_+ \\
&\subseteq \lambda(T^{-1}F)\left(\begin{pmatrix} A_1 & 0 \\ 0 & A_2 \end{pmatrix} + \begin{pmatrix} 0 & B_1 \\ \lambda^{-1}B_2 & 0 \end{pmatrix}\right)H^{(n)}_+ \\
&\subseteq \lambda T^{-1}F(\lambda^{-1}(\mathbf{C}^k)^\perp \oplus H^{(n)}_+) \\
&= T^{-1}F((\mathbf{C}^k)^\perp \oplus \lambda H^{(n)}_+) \\
&= W_1.
\end{aligned}$$

A similar calculation shows that $(W_1)_z \subseteq W_0$, and $(W_i)_{\bar{z}} \subseteq W_i$ for $i = 0, 1, 2$.

Conversely, assume that $W = FH_+^{(n)}$ satisfies (F1), with $F : \mathbf{C} \to \Lambda U_n$. Then we have

$$F^{-1}F_z = A + \tfrac{1}{\lambda}B, \quad F^{-1}F_{\bar{z}} = C + \lambda D$$

by the same argument as in the un-twisted case (i.e., by considering only W_2 and W_0). The condition $(W_2)_z \subseteq W_1$ implies that $\operatorname{Im} B \subseteq (\mathbf{C}^k)^\perp$, and the condition $(W_1)_z \subseteq W_0$ implies that $(\mathbf{C}^k)^\perp \subseteq \operatorname{Ker} B$. Hence B is of the form

$$B = \begin{pmatrix} 0 & 0 \\ * & 0 \end{pmatrix}.$$

The condition $(W_1)_{\bar{z}} \subseteq W_1$ implies that C preserves $(\mathbf{C}^k)^\perp$, i.e., C is of the form

$$C = \begin{pmatrix} * & 0 \\ * & * \end{pmatrix}.$$

Finally, using the fact that $C = -A^*$ and $D = -B^*$, we see that A, B, C, D have the required special form. ∎

As in the un-twisted case, we can use the natural action of $\Lambda GL_n\mathbf{C}$ on $\mathrm{Fl}_{n-k,k}^{(n)}$ to obtain a new symmetry group of the harmonic map equation. For $\gamma \in \Lambda GL_n\mathbf{C}$, and a solution $\{W_i\}$ of (F1), we define

$$\gamma \cdot \{W_i\} = \{\gamma W_i\}.$$

This is obviously a solution of (F1). As in the un-twisted case, one can show that the new action is determined by the old action, and that in the case of harmonic maps of finite uniton number, the new action determines the old action.

Example:

Let us return one more time to the example of a holomorphic map $f : \mathbf{C} \to \mathrm{Gr}_k(\mathbf{C}^n)$. We have seen that a corresponding solution of (Λ_σ) is

$$F(z, \lambda) = (P_k + \tfrac{1}{\lambda}P_k{}^\perp)\psi(z)(P_k + \lambda P_k{}^\perp)$$

where $P_k : \mathbf{C}^n \to \mathbf{C}^k$ is Hermitian projection and $f(z) = \psi(z)\mathbf{C}^k$. This in turn corresponds to the very simple map

$$T^{-1}F(z, \lambda) = \psi(z)$$

if we modify our definition of T slightly, so that

$$T(\gamma)(\lambda) = (P_k + \lambda P_k{}^\perp)^{-1}\gamma(\lambda^2)(P_k + \lambda P_k{}^\perp).$$

(The earlier definition of T would be appropriate if we were to consider $\psi(z)(\mathbf{C}^{n-k})^{\perp}$ instead of $\psi(z)\mathbf{C}^k$.) And this corresponds to the solution $\{W_i\}$ of (Fl) given by

$$W_2(z) = \lambda H_+^{(n)}, \quad W_1(z) = \psi(z)\mathbf{C}^k \oplus \lambda H_+^{(n)} = f(z) \oplus \lambda H_+^{(n)}, \quad W_0 = H_+^{(n)},$$

which "is" just f itself.

In the next chapter we consider a more interesting example.

Bibliographical comments.

See the comments at the end of Chapter 20.

Chapter 20: Application: Harmonic maps from S^2 to $\mathbf{C}P^n$

We give two applications of the preceding theory.

I. Harmonic maps $S^2 \to \mathbf{C}P^n$.

We have seen that any holomorphic map $f : \mathbf{C} \to \mathbf{C}P^n$ is harmonic. The following well known construction produces further examples $f = f_0, f_1, f_2, \ldots$ of harmonic maps in this situation. Let us write

$$f = [\tilde{f}], \quad \tilde{f} : \mathbf{C} \to \mathbf{C}^{n+1} - \{0\}$$

where $\tilde{f}$ is a holomorphic vector-valued function. Then we define

$$f_i = [\mathrm{Span}\{\tilde{f}, \tilde{f}', \ldots, \tilde{f}^{(i-1)}\}^\perp \cap \mathrm{Span}\{\tilde{f}, \tilde{f}', \ldots, \tilde{f}^{(i)}\}],$$

where $f^{(i)} = \partial^i f / \partial z^i$. (If $\mathrm{Span}\{\tilde{f}, \tilde{f}', \ldots, \tilde{f}^{(r)}\}$ is a constant subspace, independent of z, then f_i is undefined for $i \geq r + 1$. If $i \leq r$, then f_i is a well defined map of $\mathbf{C}$, except possibly for isolated removable singularities. We denote by f_i the map obtained by removing these singularities.)

Theorem (Eells and Wood [1983]).

(1) Each map f_i is harmonic.

(2) Conversely, if $\phi : S^2 \to \mathbf{C}P^n$ is any harmonic map, then $\phi|_\mathbf{C} = f_i$ for some holomorphic map f and some i.

The survey article of Eells and Lemaire [1988] contains some history and references to other versions of this result, which was first discovered by mathematical physicists.

We give a proof of this theorem shortly. First, however, we convert the above description of f_i into loop theoretic terms. We take

$$G = U_{n+1}, \quad G^c = GL_{n+1}\mathbf{C}$$

and

$$\sigma(X) = (P_{i-1} - P_{i-1}{}^\perp P_i + P_i{}^\perp)X(P_{i-1} - P_{i-1}{}^\perp P_i + P_i{}^\perp)^{-1}$$

where P_i now means Hermitian projection onto $\mathbf{C}^{i+1} = \mathrm{Span}\{e_0, \ldots, e_i\} = V_0 \oplus \cdots \oplus V_i$.

A solution of (Ω) corresponding to the composition of f_i with the totally geodesic embedding $L \mapsto \pi_L{}^\perp - \pi_L$ is given by

$$F(\,, \lambda) = (\pi_{i-1} + \lambda \pi_{i-1}{}^\perp)(\pi_i + \lambda \pi_i{}^\perp)$$
$$= \pi_{i-1} + \lambda \pi_{i-1}{}^\perp \pi_i + \lambda^2 \pi_i{}^\perp$$

where π_i denotes Hermitian projection onto $\mathrm{Span}\{f_0, \ldots, f_i\}$. The factorization into two linear factors is in accordance with Theorem A of Chapter 17. In this particular example, both factors are unitons, i.e., π_{i-1} and π_i are holomorphic maps into $\mathrm{Gr}_i(\mathbf{C}^{n+1})$, $\mathrm{Gr}_{i+1}(\mathbf{C}^{n+1})$ respectively.

A corresponding solution of (Gr) is

$$W = (\pi_{i-1} + \lambda \pi_{i-1}^{\perp}\pi_i + \lambda^2 \pi_i^{\perp})H_{+}^{(n+1)} = \pi_{i-1} \oplus \lambda\pi_i \oplus \lambda^2 H_{+}^{(n+1)}.$$

Note that W may be identified with the map

$$(\pi_{i-1}, \pi_i) : \mathbf{C} \to \mathrm{F}_{i,i+1}(\mathbf{C}^{n+1})$$

into the flag manifold $\mathrm{F}_{i,i+1}(\mathbf{C}^{n+1}) = GL_{n+1}\mathbf{C}(\mathbf{C}^i \subseteq \mathbf{C}^{i+1} \subseteq \mathbf{C}^{n+1})$. Conversely (see Segal [1989]), it is possible to show that if $(E_i, E_{i+1}) : \mathbf{C} \to \mathrm{F}_{i,i+1}(\mathbf{C}^{n+1})$ is a holomorphic map such that $\frac{\partial}{\partial z}E_{i-1} \subseteq E_i$, then $(E_i, E_{i+1}) = (\pi_{i-1}, \pi_i)$ for some f.

It may be verified that a solution of (Ω_σ) corresponding to f_i is

$$(P_{i-1} + \lambda P_{i-1}^{\perp}P_i + \lambda^2 P_i^{\perp})^{-1}(\pi_{i-1} + \lambda \pi_{i-1}^{\perp}\pi_i + \lambda^2 \pi_i^{\perp}).$$

To describe a solution of (Λ_σ), we choose a map $\psi : \mathbf{C} \to U_{n+1}$ such that $\pi_i = \psi P_i$ for all i. For example, we may choose

$$\psi = \begin{pmatrix} | & | & & | \\ \tilde{f} & \tilde{f}' & \cdots & \tilde{f}^{(n)} \\ | & | & & | \end{pmatrix}_u$$

where the notation X_u means the first factor in the factorization of X with respect to the Iwasawa decomposition $GL_{n+1}\mathbf{C} = U_{n+1}\hat{N}_{n+1}$. After performing the factorization (i.e., the Gram-Schmidt orthogonalization) we obtain

$$\psi = \begin{pmatrix} | & & | \\ \tilde{f}_0 & \cdots & \tilde{f}_n \\ | & & | \end{pmatrix}$$

where $\tilde{f}_i : \mathbf{C} \to \mathbf{C}^{n+1} - \{0\}$ is such that $f_i = [\tilde{f}_i]$ and $\|\tilde{f}_i\| = 1$. We assume here that $\tilde{f}_0(z), \ldots, \tilde{f}_n(z)$ are linearly independent for all z, after removal of isolated points of linear dependence. This will be referred to as the *non-degenerate case*. In the degenerate case, i.e., where $\tilde{f}_0(z), \ldots, \tilde{f}_n(z)$ span (at most) an $(m + 1)$-dimensional subspace for $m < n$, we may reduce to the non-degenerate case for the smaller group U_{m+1}. Therefore, it suffices to consider the non-degenerate case.

In terms of ψ, the above solution of (Ω_σ) is

$$(P_{i-1} + \lambda P_{i-1}^{\perp}P_i + \lambda^2 P_i^{\perp})^{-1}\psi(P_{i-1} + \lambda P_{i-1}^{\perp}P_i + \lambda^2 P_i^{\perp})\psi^{-1},$$

and the required solution of (Λ_σ) is

$$F(z,\lambda) = (P_{i-1} + \lambda P_{i-1}{}^\perp P_i + \lambda^2 P_i{}^\perp)^{-1}\psi(z)(P_{i-1} + \lambda P_{i-1}{}^\perp P_i + \lambda^2 P_i{}^\perp).$$

To obtain a corresponding solution of (Fl), we must apply $T^{-1}F$ to the basepoint $\lambda H_+^{(n+1)} \subseteq V_i \oplus H_+^{(n+1)} \subseteq H_+^{(n+1)}$ of $\mathrm{Fl}_{1,n}^{(n+1)}$, where $T : \Lambda GL_{n+1}\mathbf{C} \to (\Lambda GL_{n+1}\mathbf{C})_\sigma$ is the isomorphism

$$T(\gamma)(\lambda) = (P_{i-1} + \lambda P_{i-1}{}^\perp P_i + P_i{}^\perp)\gamma(\lambda^2)(P_{i-1} + \lambda P_{i-1}{}^\perp P_i + P_i{}^\perp)^{-1}.$$

We have

$$T^{-1}F(z,\lambda) = (P_{i-1} + \lambda P_{i-1}^\perp)^{-1}\psi(z)(P_{i-1} + \lambda P_{i-1}^\perp).$$

Applying this to the basepoint of $\mathrm{Fl}_{1,n}^{(n+1)}$, and omitting the unimportant first factor $P_{i-1} + \lambda^{-1}P_{i-1}^\perp$, we obtain $W_2 \subseteq W_1 \subseteq W_0$ with

$$W_2 = \lambda(\psi V_0 \oplus \cdots \oplus \psi V_{i-1}) \oplus \lambda^2 H_+^{(n+1)} = \lambda\pi_{i-1} \oplus \lambda^2 H_+^{(n+1)}$$

$$W_1 = \lambda(\psi V_0 \oplus \cdots \oplus \psi V_i) \oplus \lambda^2 H_+^{(n+1)} = \lambda\pi_i \oplus \lambda^2 H_+^{(n+1)}$$

$$W_0 = \psi V_0 \oplus \cdots \oplus \psi V_{i-1} \oplus \lambda H_+^{(n+1)} = \pi_{i-1} \oplus \lambda H_+^{(n+1)}.$$

This is the desired solution of (Fl).

In summary, we have found the following solutions (of the indicated systems) corresponding to the harmonic map f_i:

(Ω_σ) $\quad (P_{i-1} + \lambda P_{i-1}{}^\perp P_i + \lambda^2 P_i{}^\perp)^{-1}(\pi_{i-1} + \lambda\pi_{i-1}{}^\perp\pi_i + \lambda^2\pi_i{}^\perp)$

(Λ_σ) $\quad (P_{i-1} + \lambda P_{i-1}{}^\perp P_i + \lambda^2 P_i{}^\perp)^{-1}\psi(P_{i-1} + \lambda P_{i-1}{}^\perp P_i + \lambda^2 P_i{}^\perp)$

(Gr) $\quad\quad\quad\quad\quad \pi_{i-1} \oplus \lambda\pi_i \oplus \lambda^2 H_+^{(n+1)}$

(Fl) $\quad \lambda\pi_{i-1} \oplus \lambda^2 H_+^{(n+1)} \subseteq \lambda\pi_i \oplus \lambda^2 H_+^{(n+1)} \subseteq \pi_{i-1} \oplus \lambda H_+^{(n+1)}.$

It is an elementary matter to verify that all these formulae are correct. The origin of the formulae is perhaps less obvious. However, they do arise naturally from the "Grassmannian theoretic" proof of the theorem that we are going to give next.

Of all these equivalent formulae, the simplest are the ones for (Gr) and (Fl). And of these two, the one for (Fl) is more transparent, in the sense that it is closer to the original definition of f_i. For this reason, our proof will be expressed in terms of (Fl).

Proof of the theorem. Statement (1) follows immediately from the preceding discussion.

To prove statement (2), let us begin with a harmonic map $\phi : S^2 \to \mathbf{C}P^n$. We regard $\mathbf{C}P^n$ as $U_n/(U_n)_\sigma$, where

$$\sigma = \text{ conjugation by } \begin{pmatrix} I_n & 0 \\ 0 & -1 \end{pmatrix}.$$

We regard $\mathrm{Fl}_{1,n}^{(n+1)}$ as the orbit under $\Lambda GL_{n+1}\mathbf{C}$ of the point $\lambda H_+^{(n+1)} \subseteq V_n \oplus H_+^{(n+1)} \subseteq H_+^{(n+1)}$; the isotropy subgroup at this point is $\Lambda_+^P GL_{n+1}\mathbf{C}$, where $P = \{X \in GL_{n+1}\mathbf{C} \mid X(V_n) \subseteq V_n\}$.

By Theorem A or B of Chapter 17, there exists a corresponding solution $F : \mathbf{C} \to (\Lambda^{\text{alg}} U_{n+1})_\sigma$ of (Λ_σ). By the discussion in section II of Chapter 19 we obtain a solution $\{W_i\} : \mathbf{C} \to \mathrm{Fl}_{1,n}^{(n+1),\text{alg}}$ of (Fl).

We now make use of the Bruhat decomposition

$$\Lambda^{\text{alg}} GL_{n+1}\mathbf{C} = \bigcup_{k_0 \geq \cdots \geq k_n} \Lambda_+^{\text{alg}} GL_{n+1}\mathbf{C} \begin{pmatrix} \lambda^{k_0} & & & \\ & \lambda^{k_1} & & \\ & & \ddots & \\ & & & \lambda^{k_n} \end{pmatrix} \Lambda_+^{\text{alg}} GL_{n+1}\mathbf{C}$$

(see section IV of Chapter 17). In addition we use the decomposition

$$\Lambda_+^{\text{alg}} GL_{n+1}\mathbf{C} = \bigcup_{s \in \Sigma_{n+1}} \Delta\, s\, \Lambda_+^{P,\text{alg}} GL_{n+1}\mathbf{C}$$

which follows directly from the finite dimensional Bruhat decomposition $GL_{n+1}\mathbf{C} = \cup_{s \in \Sigma_{n+1}} \Delta s P$ (see section III of Chapter 14). Combining these, we obtain a Bruhat decomposition of $\mathrm{Fl}_{1,n}^{(n+1),\text{alg}}$:

$$\bigcup_{\substack{k_0 \geq \cdots \geq k_n \\ s \in \Sigma_{n+1}}} \Lambda_+^{\text{alg}} GL_{n+1}\mathbf{C} \begin{pmatrix} \lambda^{k_0} & & & \\ & \lambda^{k_1} & & \\ & & \ddots & \\ & & & \lambda^{k_n} \end{pmatrix} \Delta\, s\, \{V_n\},$$

where (in this proof only) $\{V_n\}$ stands for the basepoint of $\mathrm{Fl}_{1,n}^{(n+1),\text{alg}}$, i.e., $(\lambda H_+^{(n+1)} \subseteq V_n \oplus \lambda H_+^{(n+1)} \subseteq H_+^{(n+1)})$.

This is a decomposition of $\mathrm{Fl}_{1,n}^{(n+1),\text{alg}}$ into a disjoint union of finite dimensional complex manifolds. We can simplify the description somewhat by using the fact that

$$\begin{pmatrix} \lambda^{k_0} & & & \\ & \lambda^{k_1} & & \\ & & \ddots & \\ & & & \lambda^{k_n} \end{pmatrix} \Delta \begin{pmatrix} \lambda^{k_0} & & & \\ & \lambda^{k_1} & & \\ & & \ddots & \\ & & & \lambda^{k_n} \end{pmatrix}^{-1} \subseteq \Lambda_+^{\text{alg}} GL_{n+1}\mathbf{C},$$

to obtain

$$\bigcup_{\substack{k_0 \geq \cdots \geq k_n \\ j \in \{0,\ldots,n\}}} \Lambda_+^{\mathrm{alg}} GL_{n+1}\mathbf{C} \begin{pmatrix} \lambda^{k_0} & & & \\ & \lambda^{k_1} & & \\ & & \ddots & \\ & & & \lambda^{k_n} \end{pmatrix} \{V_j\}.$$

Since our map $\{W_i\} : \mathbf{C} \to \mathrm{Fl}_{1,n}^{(n+1),\mathrm{alg}}$ is holomorphic, its image must lie inside one of the above complex submanifolds, except possibly for a discrete set of points of $\mathbf{C}$. Therefore, for some $k_0, \ldots, k_n$ and some j, we must have

$$\{W_i\} = \sum_{s \geq 0} A_s \lambda^s \begin{pmatrix} \lambda^{k_0} & & & \\ & \lambda^{k_1} & & \\ & & \ddots & \\ & & & \lambda^{k_n} \end{pmatrix} \{V_j\}.$$

There is a small technical point here: We have changed to the new basepoint $\lambda H_+^{(n+1)} \subseteq V_j \oplus H_+^{(n+1)} \subseteq H_+^{(n+1)}$. So, from now on, we must redefine σ to be conjugation by $P_{j-1} - P_{j-1}{}^\perp P_j + P_j{}^\perp$.

Next we use the fact that if $\{W_i\}$ is a solution of (Fl), then we obtain another solution $\{W_i^t\}$ for any $t \in \mathbf{R}$, by replacing λ by $e^{-t}\lambda$. (This is a special case of the "dressing action". See section IV of Chapter 17.) Moreover, the same is true in the limit $t \to \infty$, i.e., we obtain a further solution $\{W_i^\infty\}$. (At worst a discrete set of removable singularities are introduced in the limit.) Applying this observation to our formula, we find that

$$\{W_i^\infty\} = A_0 \begin{pmatrix} \lambda^{k_0} & & & \\ & \lambda^{k_1} & & \\ & & \ddots & \\ & & & \lambda^{k_n} \end{pmatrix} \{V_j\}$$

is a solution of (Fl). The function A_0 takes values in $GL_{n+1}\mathbf{C}$, but we may replace it by $\psi = (A_0)_u \in U_{n+1}$ without changing $\{W_i^\infty\}$. We do this from now on in order to simplify notation. It will also simplify our argument if we assume from now on that none of the maps $\psi V_{i_1} \oplus \cdots \oplus \psi V_{i_r}$ is constant, for $V_{i_1} \oplus \cdots \oplus V_{i_r} \neq \mathbf{C}^{n+1}$. As in the example discussed earlier, we refer to this as the *non-degenerate case*. The remaining (degenerate) cases all reduce to non-degenerate situations for smaller unitary groups.

Lemma. *In the non-degenerate case, if*

$$W = \psi \begin{pmatrix} \lambda^{k_0} & & & \\ & \lambda^{k_1} & & \\ & & \ddots & \\ & & & \lambda^{k_n} \end{pmatrix} H_+^{(n+1)}$$

is a solution of (Gr) *with $k_0 \geq \cdots \geq k_n$, then we must have $k_{i-1} - k_i = 0$ or 1 for all i.*

Proof. We have

$$W = \lambda^{k_n} \psi V_n \oplus \lambda^{k_n+1} \psi V_n \oplus \cdots \oplus \lambda^{k_{n-1}-1} \psi V_n \oplus \lambda^{k_n}(\psi V_n \oplus \psi V_{n-1}) \oplus \cdots.$$

The condition $(\lambda W)_z \subseteq W$ of (Gr) says that

$$\lambda^{k_n+1}(\psi V_n)_z \oplus \cdots \oplus \lambda^{k_{n-1}}(\psi V_n)_z \oplus \lambda^{k_n+1}((\psi V_n)_z \oplus (\psi V_{n-1})_z) \oplus \cdots$$
$$\subseteq \lambda^{k_n} \psi V_n \oplus \lambda^{k_n+1} \psi V_n \oplus \cdots \oplus \lambda^{k_{n-1}-1} \psi V_n \oplus \lambda^{k_n}(\psi V_n \oplus \psi V_{n-1}) \oplus \cdots.$$

If $k_{n-1} - k_n \geq 2$, this implies that $(\psi V_n)_z \subseteq \psi V_n$. But from the condition $W_{\bar{z}} \subseteq W$ of (Gr) we see that $(\psi V_n)_{\bar{z}} \subseteq \psi V_n$. Hence ψV_n is both holomorphic and anti-holomorphic, and therefore constant. In the non-degenerate case this is impossible, so we conclude that $k_{n-1} - k_n = 0$ or 1. A similar argument applies to $k_{i-1} - k_i$ for all i. ∎

The lemma allows us to simplify $k_0, \ldots, k_n$. But it only makes use of (Gr), and we can make a further simplification by using the full condition (Fl). To do this we apply the argument of the lemma to

$$(W_2 \subseteq W_1) = \psi \begin{pmatrix} \lambda^{k_0} & & & \\ & \lambda^{k_1} & & \\ & & \ddots & \\ & & & \lambda^{k_n} \end{pmatrix} (\lambda H_+^{(n+1)} \subseteq V_j \oplus \lambda H_+^{(n+1)})$$

where we now assume (by the lemma) that

$$k_0 = \cdots = k_{r_0} = k$$
$$k_{r_0+1} = \cdots = k_{r_1} = k - 1$$
$$\cdots \quad \cdots$$
$$k_{r_{s-1}+1} = \cdots = k_{r_s} = k_n = 1$$

for some $r_0, \ldots, r_s = n$. (We may assume that $k_n = 1$ by multiplying by the scalar λ^{-k_n}.)

Step 1: First we consider the terms in λ in the above expression for "$W_2 \subseteq W_1$". They are:

$$\lambda(\psi V_{r_s} \oplus \cdots \oplus \psi V_{r_{s-1}+1}) \subseteq \begin{cases} \lambda(\psi V_{r_s} \oplus \cdots \oplus \psi V_{r_{s-1}+1}) \\ \quad \text{if } j \notin \{r_{s-2}+1, \ldots, r_{s-1}\} \\ \lambda(\psi V_{r_s} \oplus \cdots \oplus \psi V_{r_{s-1}+1}) \oplus \lambda \psi V_j \\ \quad \text{if } j \in \{r_{s-2}+1, \ldots, r_{s-1}\}. \end{cases}$$

From (Fl) we have $(W_2)_z \subseteq W_1$, which implies (for the terms in λ)

$$(\psi V_{r_s} \oplus \cdots \oplus \psi V_{r_{s-1}+1})_z \subseteq \begin{cases} \psi V_{r_s} \oplus \cdots \oplus \psi V_{r_{s-1}+1} \\ \quad \text{if } j \notin \{r_{s-2}+1, \ldots, r_{s-1}\} \\ (\psi V_{r_s} \oplus \cdots \oplus \psi V_{r_{s-1}+1}) \oplus \psi V_j \\ \quad \text{if } j \in \{r_{s-2}+1, \ldots, r_{s-1}\}. \end{cases}$$

In the non-degenerate case, only the second possibility can hold, so we conclude that $j \in \{r_{s-2}+1, \ldots, r_{s-1}\}$.

Step 2: Next we consider the terms in λ^2:

$$\lambda^2(\psi V_{r_s} \oplus \cdots \oplus \psi V_{r_{s-2}+1}) \subseteq \lambda^2(\psi V_{r_s} \oplus \cdots \oplus \psi V_{r_{s-2}+1})$$

(using the fact that $j \in \{r_{s-2}+1, \ldots, r_{s-1}\}$). Here, (Fl) gives

$$(\psi V_{r_s} \oplus \cdots \oplus \psi V_{r_{s-2}+1})_z \subseteq \psi V_{r_s} \oplus \cdots \oplus \psi V_{r_{s-2}+1},$$

which is impossible in the non-degenerate case. We conclude that $s = 1$.

We have therefore reached the form

$$\{W_i^\infty\} = \psi \begin{pmatrix} \lambda & & & & & \\ & \ddots & & & & \\ & & \lambda & & & \\ & & & 1 & & \\ & & & & \ddots & \\ & & & & & 1 \end{pmatrix} \{V_j\}$$

where the first $r_0 + 1$ entries of the diagonal matrix are λ, and $0 \le j \le r_0$. Going back to our original solution, we deduce that

$$\{W_i\} = \sum_{s \ge 0} A_s \lambda^s \begin{pmatrix} \lambda & & & & & \\ & \ddots & & & & \\ & & \lambda & & & \\ & & & 1 & & \\ & & & & \ddots & \\ & & & & & 1 \end{pmatrix} \{V_j\}.$$

By an obvious extension of the theorem from section II of Chapter 17, it follows that the above expressions for $\{W_i^\infty\}$ and $\{W_i\}$ are actually *equal*. Thus we have

$$W_2 = \lambda(\psi V_{r_0+1} \oplus \cdots \oplus \psi V_n) \oplus \lambda^2 H_+^{(n+1)}$$
$$W_1 = \lambda(\psi V_j \oplus (\psi V_{r_0+1} \oplus \cdots \oplus \psi V_n)) \oplus \lambda^2 H_+^{(n+1)}$$
$$W_0 = \psi V_{r_0+1} \oplus \cdots \oplus \psi V_n \oplus \lambda H_+^{(n+1)}.$$

So far we have shown only that any solution of (Fl) must have this form. However, it is clear that $\{W_i\}$ always satisfies (Fl), whenever ψ has the property

$$(\psi V_n \oplus \cdots \oplus \psi V_{r_0+1})_z \subseteq (\psi V_n \oplus \cdots \oplus \psi V_{r_0+1}) \oplus \psi V_j.$$

Indeed, apart from minor changes of notation, $\{W_i\}$ agrees with the solution of (Fl) given after the statement of the theorem. We have therefore found the most general harmonic map from S^2 to $\mathbf{CP}^n$. ∎

II. Estimates of the uniton number.

In Chapter 17 we mentioned Uhlenbeck's result that the minimal uniton number of a harmonic map $\phi : S^2 \to U_n$ is at most $n-1$. The lemma of the previous section gives an independent proof of this fact, as it shows that any (non-degenerate) solution of (Gr) must be of the form

$$W = \sum_{i\geq 0} A_i \lambda^i \begin{pmatrix} \lambda^{k_1} & & \\ & \ddots & \\ & & \lambda^{k_n} \end{pmatrix} H_+^{(n)}$$

with

$$k_1 = \cdots = k_{r_0} = k$$
$$k_{r_0+1} = \cdots = k_{r_1} = k-1$$
$$\cdots \quad \cdots$$
$$k_{r_{s-1}+1} = \cdots = k_{r_s} = k_n = 1.$$

We claim that the uniton number of this W is k. (Since $k \leq n-1$, this will complete the proof.) It suffices to show that $\lambda^k H_+^{(n)} \subseteq W$. Evidently we have

$$\begin{pmatrix} \lambda^{k_1} & & \\ & \ddots & \\ & & \lambda^{k_n} \end{pmatrix}^{-1} H_+^{(n)} \subseteq \lambda^{-k} H_+^{(n)}.$$

If we now apply

$$\sum_{i\geq 0} A_i \lambda^i \begin{pmatrix} \lambda^{k_1} & & \\ & \ddots & \\ & & \lambda^{k_n} \end{pmatrix} \lambda^k$$

to both sides, we obtain $\lambda^k H_+^{(n)} \subseteq W$, as required.

The theorem of the previous section shows that a *stronger* result holds when ϕ factors through $\mathbf{CP}^{n-1}$, namely that the minimal uniton number is at most 2.

For maps which factor through $\text{Gr}_k(\mathbf{C}^n)$, the situation is as follows:

Theorem. *The minimal uniton number of a harmonic map from S^2 into $\mathrm{Gr}_k(\mathbf{C}^n)$ is at most $2\min\{k, n-k\}$.*

Proof. We modify the proof of the theorem of the previous section, replacing V_j by $V_{i_1} \oplus \cdots \oplus V_{i_k}$. Without loss of generality we may assume that $k \leq n/2$. The new feature is that the two step analysis of $W_2 \subseteq W_1$ must now be replaced by (at worst) a k step analysis. Let us consider this in more detail:

Step 1: We obtain the condition

$$\lambda(\psi V_{r_s} \oplus \ldots \oplus \psi V_{r_{s-1}+1})_z$$
$$\subseteq \lambda(\psi V_{r_s} \oplus \cdots \oplus \psi V_{r_{s-1}+1}) \oplus \lambda(\oplus_{i_j \in \{r_{s-2}+1,\ldots,r_{s-1}\}} \psi V_{i_j})$$

where $\psi = (A_0)_u$. By the non-degeneracy assumption, the set

$$S_1 = \{j \mid r_{s-2} + 1 \leq i_j \leq r_{s-1}\}$$

is non-empty.

Step 2: Next we obtain the condition

$$\lambda^2(\psi V_{r_s} \oplus \ldots \oplus \psi V_{r_{s-2}+1})_z$$
$$\subseteq \lambda^2(\psi V_{r_s} \oplus \cdots \oplus \psi V_{r_{s-2}+1}) \oplus \lambda^2(\oplus_{i_j \in \{r_{s-3}+1,\ldots,r_{s-2}\}} \psi V_{i_j}).$$

If $S_1 = \{1, \ldots, k\}$, then we are in the situation of the previous section, and the uniton number is at most 2. Otherwise, by the non-degeneracy assumption, the set

$$S_2 = \{j \mid r_{s-3} + 1 \leq i_j \leq r_{s-2}\}$$

is non-empty.

If $S_1 \cup S_2 = \{1, \ldots, k\}$, then we conclude that our solution of (Fl) is of the form

$$\sum_{i \geq 0} A_i \lambda^i \begin{pmatrix} \lambda^2 & & & & & & \\ & \ddots & & & & & \\ & & \lambda^2 & & & & \\ & & & \lambda & & & \\ & & & & \ddots & & \\ & & & & & \lambda & \\ & & & & & & 1 \\ & & & & & & & \ddots \\ & & & & & & & & 1 \end{pmatrix} \{V_{i_1} \oplus \cdots \oplus V_{i_k}\}$$

where $\{V_{i_1} \oplus \cdots \oplus V_{i_k}\}$ denotes the basepoint ($\lambda H_+^{(n)} \subseteq V_{i_1} \oplus \cdots \oplus V_{i_k} \oplus \lambda H_+^{(n)} \subseteq H_+^{(n)}$). Applying T^{-1}, we see that the uniton number is at most 4.

If $S_1 \cup S_2 \neq \{1, \ldots, k\}$, then we go on to Step 3.

This process terminates after at most k steps with $S_1 \cup \cdots \cup S_k = \{1, \ldots, k\}$; we conclude that the uniton number is at most $2k$. ∎

Bibliographical comments for Chapters 18-20.

The fact that harmonic maps into symmetric spaces correspond to solutions of (Λ_σ) was observed by Rawnsley (cf. Rawnsley [1988]), following Uhlenbeck's treatment (in §8 of Uhlenbeck [1989]) for the case $G/K = \mathrm{Gr}_k(\mathbf{C}^n)$. Our proof in Chapter 18 follows the one given in Burstall and Pedit [1994]. In that paper, a map ψ such that $\phi = [\psi]$ is called a *framing* of ϕ, and a solution of (Λ_σ) is called an *extended framing*. In §6 of Guest and Ohnita [1994], the version of this result for (Ω_σ) is given.

Our approach in Chapter 19 to harmonic maps into symmetric spaces via (Fl) seems to be new. A similar approach was given by Segal [1989], though he made use of the inclusion of $(\Lambda GL_n\mathbf{C})_\sigma/(\Lambda_+ GL_n\mathbf{C})_\sigma$ in the Grassmannian $\Lambda GL_n\mathbf{C}/\Lambda_+ GL_n\mathbf{C}$, rather than the identification with the flag manifold $\Lambda GL_n\mathbf{C}/\Lambda_+^P GL_n\mathbf{C}$.

For the applications in Chapter 20, the basic technique is the "deformation" $\{W^t\}$ of a solution W of (Fl). This deformation is simply the result of applying the gradient flow of a Morse function to the image of W (see section IV of Chapter 17). For further applications of such deformations, see Guest and Ohnita [1993]; Furuta *et al.* [1994].

Our result[5] on the minimal uniton number of a harmonic map $S^2 \to \mathrm{Gr}_k(\mathbf{C}^n)$ was conjectured in Uhlenbeck [1989] (Problem 9). More generally, estimates for the minimal uniton number of a harmonic map from S^2 into any compact Lie group or compact (inner) symmetric space are given in Burstall and Guest [preprint].

---

[5]This result has also been obtained by Y.-X. Dong and Y.-B. Shen, "On factorization theorems of harmonic maps into $U(N)$ and minimal uniton numbers", Hangzhou University preprint.

Chapter 21: Primitive maps

I. Primitive maps into k-symmetric spaces.

Let $\sigma : G \to G$ be an automorphism of order k. It is natural to ask whether the theory of Chapters 18 and 19 (for the case $k = 2$) can be extended to the case of general k. The answer to this question is: "not exactly", as we explain next.

First, if we replace "involution" by "automorphism of order k" in the definition of a symmetric space, then we obtain the definition of a k-*symmetric space*.

Next, let us consider the system (Λ_σ), where σ is an automorphism of order k, and $k > 2$. If $F : \mathbf{C} \to (\Lambda G)_\sigma$ is a solution of (Λ_σ), let us consider the map

$$\phi : \mathbf{C} \to G/K, \quad \phi(z) = [\psi(z)], \quad \text{where } \psi(z) = F(z, 1).$$

It can be shown that ϕ is harmonic (with respect to a certain metric on G/K), but this is not very interesting because ϕ is an extremely special harmonic map. A general harmonic map from $\mathbf{C}$ to G/K certainly does not arise from such an F.

On the other hand, if we are prepared to forget about harmonic maps for the moment, then we find that the map ϕ has a particularly simple characterization. To describe this characterization, we make use of the eigenspace decomposition

$$\mathbf{g} \otimes \mathbf{C} = \bigoplus_{i=0}^{k-1} \mathbf{g}_i$$

introduced in Chapter 13. Thus, $\mathbf{g}_i = \{ X \in \mathbf{g} \otimes \mathbf{C} \mid \sigma(X) = \omega^i X \}$, where $\sigma : \mathbf{g} \otimes \mathbf{C} \to \mathbf{g} \otimes \mathbf{C}$ denotes the complexified derivative of $\sigma : G \to G$, and $\omega = e^{2\pi\sqrt{-1}/k}$. We extend this notation by defining $\mathbf{g}_{i+kn} = \mathbf{g}_i$, for any $n \in \mathbf{Z}$. Observe that $\mathbf{g}_0 = \mathbf{k} \otimes \mathbf{C}$.

Let F be a solution of (Λ_σ). This means that F takes values in $(\Lambda G)_\sigma$ and

$$F^{-1} F_z = A + \tfrac{1}{\lambda} B$$
$$F^{-1} F_{\bar{z}} = C + \lambda D$$

where

$$A, C : \mathbf{C} \to \mathbf{g}_0, \ B : \mathbf{C} \to \mathbf{g}_{-1}, \ D : \mathbf{C} \to \mathbf{g}_1$$

with $A = c(C), D = c(B)$. Therefore $\psi = F(\ ,1)$ satisfies the following conditions:

(P)
$$\psi^{-1}\psi_z \in \mathbf{g}_0 \oplus \mathbf{g}_{-1}$$
$$\psi^{-1}\psi_{\bar{z}} \in \mathbf{g}_0 \oplus \mathbf{g}_1.$$

Following Burstall and Pedit [1994], we say that a map ϕ is *primitive* if it is of the form $\phi = [\psi]$ where ψ satisfies (P). Thus, a solution of (Λ_σ) gives rise to a primitive map.

Conversely, if $\phi : \mathbf{C} \to G/K$ is primitive, we claim that we can construct a solution F of (Λ_σ) such that $\phi(z) = [F(z,1)]$. We may write $\phi = [\psi]$, with

$$\psi^{-1}\psi_z = A + B$$
$$\psi^{-1}\psi_{\bar{z}} = C + D$$

where A, B, C, D are as above. We then have the zero-curvature equation

$$(A + B)_z - (C + D)_{\bar{z}} = [A + B, C + D]$$

as usual. Let us now write separately the components of this equation in each of the subspaces $\mathbf{g}_{-1}, \mathbf{g}_0, \mathbf{g}_1$. They are:

$$B_z = [B, C]$$
$$A_z - C_{\bar{z}} = [A, C] + [B, D]$$
$$-D_{\bar{z}} = [A, D]$$

(here we use the fact that $[\mathbf{g}_i, \mathbf{g}_j] \subseteq \mathbf{g}_{i+j}$).

These equations imply the following "zero-curvature equation with parameter":

$$(A + \tfrac{1}{\lambda}B)_z - (C + \lambda D)_{\bar{z}} = [A + \tfrac{1}{\lambda}B, C + \lambda D].$$

Hence, there exists a solution $F : \mathbf{C} \to (\Lambda G)_\sigma$ of the system

$$F^{-1}F_z = A + \tfrac{1}{\lambda}B$$
$$F^{-1}F_{\bar{z}} = C + \lambda D$$

i.e., a solution of the system (Λ_σ) – as required!

It is important to notice that (Λ_σ) is equivalent to (P) *only* when $k > 2$. There is no such equivalence when $k = 2$. In fact, when $k = 2$, we have $\mathbf{g}_{-1} = \mathbf{g}_1$, and the system (P) imposes no conditions on ϕ.

To summarize, we can say that *solutions of the system* (Λ_σ) *correspond to harmonic maps when* $k = 2$ *and to primitive maps when* $k > 2$. In this chapter we investigate the geometrical significance of primitive maps, and their relationship to harmonic maps and to the 2DTL.

II. Examples of primitive maps.

Let $G = U_{n+1}$ and

$$\sigma = \text{conjugation by} \begin{pmatrix} 1 & & & \\ & \omega & & \\ & & \ddots & \\ & & & \omega^n \end{pmatrix}^{-1}$$

where $\omega = e^{2\pi\sqrt{-1}/(n+1)}$. We shall discuss some examples of primitive maps $\mathbf{C} \to U_{n+1}/T_{n+1}$, where T_{n+1} – i.e., $(U_{n+1})_\sigma$ – denotes the subgroup of diagonal matrices, as usual. Recall from Chapter 3 that the flag manifold U_{n+1}/T_{n+1} may be described as

$$\{(L_0, \ldots, L_n) \mid L_0, \ldots, L_n \text{ are orthogonal lines in } \mathbf{C}^{n+1}\}.$$

Let $f_0 : \mathbf{C} \to \mathbf{C}P^n$ be a harmonic map. Let f_i be the harmonic map obtained from f_0 by the inductive formula

$$\hat{f}_i \overset{\triangle}{=} \text{component of } \partial \hat{f}_{i-1}/\partial z \text{ orthogonal to } \hat{f}_{i-1}$$

where $\hat{f}_i$ is a map from (a dense open subset of) $\mathbf{C}$ to $\mathbf{C}^{n+1}$ such that $f_i = [\hat{f}_i]$. The sequence

$$f_0, f_1, f_2, \ldots$$

is called the *harmonic sequence* of f_0. For a detailed justification of this construction we refer to Bolton and Woodward [1992]; Burstall and Wood [1986]; Eells and Wood [1983]; Wolfson [1985]. We have already considered a special case in section I of Chapter 20, namely when f_0 is holomorphic.

We consider now two special types of behaviour of this sequence.

(i) The periodic case.

Let us assume that $f_i = f_{i+n+1}$ for all i and $f_i \perp f_j$ for $i \neq j$. Following Bolton *et al.* [1995], we say that the maps f_i are *superconformal*.

Given f_0, it is possible to choose the maps $\hat{f}_i$ such that the following conditions are satisfied (on a dense open subset of $\mathbf{C}$, and after a suitable change of coordinates):

(i) $(\hat{f}_k)_z = \hat{f}_{k+1} + a_k \hat{f}_k$, where $a_k = (\log \|\hat{f}_k\|^2)_z$;

(ii) $(\hat{f}_k)_{\bar{z}} = b_{k-1}\hat{f}_{k-1}$, where $b_{k-1} = -\|\hat{f}_k\|^2/\|\hat{f}_{k-1}\|^2$;

(iii) $\hat{f}_{n+1} = \hat{f}_0$.

This is proved in Bolton *et al.* [1995].

Let us now define

$$\psi = \begin{pmatrix} | & & | \\ \hat{f}_0 & \cdots & \hat{f}_n \\ | & & | \end{pmatrix} \begin{pmatrix} ||\hat{f}_0||^{-1} & & \\ & \ddots & \\ & & ||\hat{f}_n||^{-1} \end{pmatrix}.$$

We claim that $[\psi] : \mathbf{C} \to U_{n+1}/T_{n+1}$ is a primitive map. To prove this we must calculate $\psi^{-1}\psi_z$. In order to simplify the calculation, we write

$$w_i = \log ||\hat{f}_i||.$$

It is then easy to verify that $\psi^{-1}\psi_z =$

$$\begin{pmatrix} (w_0)_z & & & \\ & (w_1)_z & & \\ & & \ddots & \\ & & \ddots & \\ & & & (w_n)_z \end{pmatrix} + \begin{pmatrix} 0 & & & & W_{0,n} \\ W_{1,0} & 0 & & & \\ & \ddots & \ddots & & \\ & & \ddots & \ddots & \\ & & & W_{n,n-1} & 0 \end{pmatrix}$$

where $W_{i,i-1} = e^{w_i - w_{i-1}}$. This shows that ψ is primitive.

In addition, we see that ψ gives a solution of the 2DTL! (It is possible to choose the maps $\hat{f}_i$ such that ψ takes values in SU_{n+1}, rather than in U_{n+1}.) Conversely, by reversing this construction, we see that a solution of the 2DTL gives a periodic sequence of harmonic maps. Thus, *this example gives a geometrical interpretation of the 2DTL.*

Given $f_0, \ldots, f_n$, there is some ambiguity in the choice of $\hat{f}_0, \ldots, \hat{f}_n$ (and hence of $w_0, \ldots, w_n$). For the details, we refer to section 4 of Bolton *et al.* [1995]. Given $w_0, \ldots, w_n$, the harmonic sequence $f_0, \ldots, f_n$ is determined uniquely.

There is a slight difference between the form of (Λ_σ) in Chapters 13-15 and Chapters 16-21, in that the roles of λ and λ^{-1} have been interchanged. This is a purely nominal difference (which we tolerate for historical reasons). We have changed the definition of σ, by replacing ω by ω^{-1}, in order to preserve the form of the 2DTL equation.

(ii) The finite case.

Let us assume that the harmonic sequence terminates after a finite number of steps, i.e., it is of the form

$$f_0, \ldots, f_r$$

for some $r \geq 0$. It follows that f_r is anti-holomorphic. It can be shown further (see Eells and Wood [1983]) that there exists a harmonic sequence

$$f_{-s}, \ldots, f_0, \ldots, f_r$$

(extending the previous sequence), where f_{-s} is holomorphic, and that $f_i \perp f_j$ for $i \neq j$. (Hence $r + s \leq n$.) Without loss of generality, therefore, it suffices to study finite harmonic sequences of the form

$$f_0, \ldots, f_n$$

with f_0 holomorphic and f_n anti-holomorphic.

In this case, we may choose $\hat{f}_i$ such that

(i) $(\hat{f}_k)_z = \hat{f}_{k+1} + a_k \hat{f}_k$, $k = 0, \ldots, n-1$, and $(\hat{f}_n)_z = a_n \hat{f}_n$,

(ii) $(\hat{f}_k)_{\bar{z}} = b_{k-1} \hat{f}_{k-1}$, $k = 1, \ldots, n$, and $(\hat{f}_0)_{\bar{z}} = 0$

where $a_k = (\log \|\hat{f}_k\|^2)_z$ and $b_{k-1} = -\|\hat{f}_k\|^2 / \|\hat{f}_{k-1}\|^2$.

If we define ψ as in the periodic case, we obtain $\psi^{-1} \psi_z =$

$$
\begin{pmatrix}
(w_0)_z & & & \\
& (w_1)_z & & \\
& & \ddots & \\
& & & (w_n)_z
\end{pmatrix}
+
\begin{pmatrix}
0 & & & \\
W_{1,0} & 0 & & \\
& \ddots & \ddots & \\
& & W_{n,n-1} & 0
\end{pmatrix}.
$$

Thus, $[\psi] : \mathbf{C} \to U_{n+1}/T_{n+1}$ is a primitive map.

This example is equivalent to the *finite open* 2DTL, i.e., the system

$$
\begin{aligned}
2(w_0)_{z\bar{z}} &= W_{1,0}^2 \\
2(w_i)_{z\bar{z}} &= W_{i+1,i}^2 - W_{i,i-1}^2 \quad i = 1, \ldots, n-1 \\
2(w_n)_{z\bar{z}} &= \qquad -W_{n,n-1}^2
\end{aligned}
$$

where $w_i : \mathbf{C} \to \mathbf{R}$ and $\sum_{i=0}^n w_i = 0$.

In this case we can construct explicitly a corresponding solution to the system (Λ_σ). Namely, we take

$$
G(z, \lambda) =
\begin{pmatrix}
1 & & & \\
& \lambda^{-1} & & \\
& & \ddots & \\
& & & \lambda^{-n}
\end{pmatrix}
\psi(z)
\begin{pmatrix}
1 & & & \\
& \lambda & & \\
& & \ddots & \\
& & & \lambda^n
\end{pmatrix}.
$$

It is clear that G takes values in $(\Lambda U_{n+1})_\sigma$, and that $G(z, 1) = \psi(z)$. We must verify that G is a solution of (Λ_σ). We have

$$
G^{-1}G_z = \begin{pmatrix} 1 & & & \\ & \lambda^{-1} & & \\ & & \cdots & \\ & & & \lambda^{-n} \end{pmatrix} \psi^{-1}\psi_z \begin{pmatrix} 1 & & & \\ & \lambda & & \\ & & \cdots & \\ & & & \lambda^{n} \end{pmatrix}
$$

$$
= \begin{pmatrix} (w_0)_z & & & \\ \lambda^{-1}W_{1,0} & (w_1)_z & & \\ & \cdots & \cdots & \\ & & \cdots & \\ & & \lambda^{-1}W_{n,n-1} & (w_n)_z \end{pmatrix}
$$

$= $ linear in $\frac{1}{\lambda}$.

A similar calculation shows that $G^{-1}G_{\bar{z}}$ is linear in λ, as required. (Note that this formula does *not* work in the periodic case!)

For maps which extend to $\mathbf{C} \cup \infty = S^2$, *every* primitive map is of this form:

Theorem. *Any primitive map* $\phi : S^2 \to F_{1,2,\ldots,n}(\mathbf{C}^{n+1})$ *arises from a finite harmonic sequence of length* $n+1$, *in the manner described above.* ∎

This may be proved by the method of Chapter 20, section I, using the Bruhat decomposition of $Fl_{1,1,\ldots,1}^{(n+1),\mathrm{alg}}$.

Remark: Examples of periodic harmonic sequences are quite difficult to find. In contrast, finite harmonic sequences, being given by holomorphic maps $\mathbf{C} \to \mathbf{C}P^n$, are plentiful.

III. From primitive maps to harmonic maps: First method.

If F is a solution of (Λ_σ), then there is an obvious way to obtain a harmonic map. Namely, we have the inclusion

$$(\Lambda G)_\sigma \subseteq \Lambda G,$$

so we may regard F as a solution of (Λ). By the theory of Chapter 16, this gives the harmonic map $\phi = F(\ ,-1)F(\ ,1)^{-1} : \mathbf{C} \to G$.

More generally, if $\tau : G \to G$ is an involution such that

$$(\Lambda G)_\sigma \subseteq (\Lambda G)_\tau,$$

then we obtain a harmonic map $\phi : \mathbf{C} \to G/G_\tau$. This map ϕ is obtained from the primitive map $[F(\ ,1)] : \mathbf{C} \to G/G_\sigma$ by composition with the natural projection map $G/G_\sigma \to G/G_\tau$.

Example:

Let $f : \mathbf{C} \to U_{n+1}/T_{n+1}$ be the primitive map constructed in (ii) of section II. Using the inclusion $(\Lambda U_{n+1})_\sigma \subseteq \Lambda U_{n+1}$, we obtain a harmonic map

$$\phi(z) = (p_0 - p_1 + \cdots + (-1)^n p_n)(\pi_0(z) - \pi_1(z) + \cdots + (-1)^n \pi_n(z))$$

where $p_i, \pi_i(z)$ denote the Hermitian projection operators onto the i-th coordinate axis and $f_i(z)$, respectively.

IV. From primitive maps to harmonic maps: Second method.

If σ is an inner automorphism of G, and if F is a solution of (Λ_σ), then there is another way to obtain a harmonic map. Namely, we use the isomorphism

$$(\Lambda G)_\sigma \cong \Lambda G$$

(mentioned in section II of Chapter 11). We shall discuss this in the case where $G = U_{n+1}$ and

$$\sigma \quad = \quad \text{conjugation by} \quad \begin{pmatrix} 1 & & & \\ & \omega & & \\ & & \ddots & \\ & & & \omega^n \end{pmatrix}^{-1}$$

where $\omega = e^{2\pi\sqrt{-1}/(n+1)}$. In this case, the isomorphism

$$T : \Lambda U_{n+1} \to (\Lambda U_{n+1})_\sigma$$

is given explicitly by

$$T(\gamma)(\lambda) = \begin{pmatrix} 1 & & & \\ & \lambda^{-1} & & \\ & & \ddots & \\ & & & \lambda^{-n} \end{pmatrix} \gamma(\lambda^{n+1}) \begin{pmatrix} 1 & & & \\ & \lambda & & \\ & & \ddots & \\ & & & \lambda^n \end{pmatrix}$$

(and the same formula applies to the Lie algebras, as well as to the complexified Lie groups and Lie algebras).

Exactly as in the lemma of section II of Chapter 19, we have:

Proposition 1. *If F is a solution of (Λ_σ), then $T^{-1}(F)$ is a solution of (Λ).*

Proof. Let $A + \lambda^{-1}B, C + \lambda D \in (\Lambda \mathbf{g} \otimes \mathbf{C})_\sigma$. We must show that $T^{-1}(A + \lambda^{-1}B)$, $T^{-1}(C + \lambda D)$ are linear in λ^{-1}, λ, respectively. We have

$$A + \lambda^{-1}B = \begin{pmatrix} a_0 & & & \\ & a_1 & & \\ & & \ddots & \\ & & & a_n \end{pmatrix} + \frac{1}{\lambda}\begin{pmatrix} 0 & & & b_0 \\ b_1 & 0 & & \\ & \cdots & \cdots & \\ & & \cdots & \cdots \\ & & b_n & 0 \end{pmatrix}.$$

It is easy to verify that

$$T^{-1}(A + \tfrac{1}{\lambda}B) = \begin{pmatrix} a_0 & & & & \lambda^{-1}b_0 \\ b_1 & a_1 & & & \\ & \cdots & \cdots & & \\ & & \cdots & \cdots & \\ & & & b_n & a_n \end{pmatrix}.$$

A similar calculation can be made for $T^{-1}(C + \lambda D)$. ∎

Next, consider an automorphism

$$\sigma' = \text{conjugation by} \begin{pmatrix} \omega^{k_0} & & & \\ & \omega^{k_1} & & \\ & & \ddots & \\ & & & \omega^{k_n} \end{pmatrix}^{-1}$$

where $0 = k_0 \leq k_1 \leq \cdots \leq k_n = k$, $k_{i+1} - k_i = 0$ or 1 for all i, and $\omega = e^{2\pi\sqrt{-1}/k}$. This has order k. There is a corresponding isomorphism

$$T' : \Lambda U_{n+1} \to (\Lambda U_{n+1})_{\sigma'}$$

which is given by

$$T'(\gamma)(\lambda) = \begin{pmatrix} \lambda^{-k_0} & & & \\ & \lambda^{-k_1} & & \\ & & \ddots & \\ & & & \lambda^{-k_n} \end{pmatrix} \gamma(\lambda^k) \begin{pmatrix} \lambda^{k_0} & & & \\ & \lambda^{k_1} & & \\ & & \ddots & \\ & & & \lambda^{k_n} \end{pmatrix}.$$

Proposition 2. *If F is a solution of (Λ_σ), then $T'T^{-1}F$ is a solution of $(\Lambda_{\sigma'})$.*

Proof. In the proof of Proposition 1, we found that

$$T^{-1}(A + \tfrac{1}{\lambda}B) = \begin{pmatrix} a_0 & & & & \lambda^{-1}b_0 \\ b_1 & a_1 & & & \\ & \cdots & \cdots & & \\ & & \cdots & \cdots & \\ & & & b_n & a_n \end{pmatrix}.$$

The result of applying T' to this is clearly linear in λ^{-1}. A similar calculation can be made for $T'T^{-1}(C + \lambda D)$. ∎

Proposition 2 establishes a "transform procedure" for primitive maps. As in the previous section, we can consider $T'T^{-1}F$ to be a solution of (Λ), via the inclusion $(\Lambda U_{n+1})_{\sigma'} \subseteq \Lambda U_{n+1}$. This is valid for any σ' of the above form. Hence, from a solution F of (Λ_σ), we obtain a collection of solutions of (Λ). In other words, *from a primitive map we obtain a collection of harmonic maps.*

The effect of the transformation $T'T^{-1}$ on a primitive map $\phi : \mathbf{C} \to G/G_\sigma$ is simply to compose with the natural projection map $G/G_\sigma \to G/G_{\sigma'}$. (Proof: The maps T, T' have no effect on the value of loops at $\lambda = 1$, so $T'T^{-1}F(z,1) = F(z,1)$.)

Example:

Let σ' be conjugation by $P_{k-1} - P_{k-1}^{\perp} P_k + P_k^{\perp}$ where P_k is the Hermitian projection operator onto $\mathbf{C}^{k+1} = \mathrm{Span}\{e_0, \ldots, e_k\}$. Let F be a solution of (Λ_σ) which corresponds to a periodic or finite harmonic sequence, as in section II. Then $T'T^{-1}F$ is a solution of $(\Lambda_{\sigma'})$. What is the resulting harmonic map into the symmetric space $U_{n+1}/(U_{n+1})_{\sigma'} \cong \mathbf{C}P^n$? By the remarks above, it is given by composing the original primitive map with the natural projection map

$$U_{n+1}/(U_{n+1})_\sigma \to U_{n+1}/(U_{n+1})_{\sigma'},$$

and so it is just the k-th harmonic map f_k. (Thus, the "columns" of a primitive map $\mathbf{C} \to U_{n+1}/T_{n+1}$ are harmonic maps $\mathbf{C} \to \mathbf{C}P^n$.)

In the case of a finite harmonic sequence, we have seen that $T^{-1}F = \psi$, and so we have the explicit formula

$$T'T^{-1}F = (P_{k-1} + \lambda P_{k-1}^{\perp} P_k + \lambda^2 P_k^{\perp})^{-1} \psi (P_{k-1} + \lambda P_{k-1}^{\perp} P_k + \lambda^2 P_k^{\perp}).$$

This is in agreement with the formula of Chapter 20 (see the beginning of section I).

Bibliographical comments.

Primitive maps into k-symmetric spaces were introduced and studied in Burstall and Pedit [1994].

The idea of passing "from primitive maps to harmonic maps" first arose in McIntosh [1994b]. This depended on earlier results in twistor theory, from Black [1991]. Our treatment in section IV is a self-contained loop theoretic version of this.

Relations between harmonic maps and solutions to the 2DTL have been studied in Bolton *et al.* [1995]; Doliwa and Sym [1994]; Fujii [1993]; Miyaoka [in press]; our treatment follows that of the first reference. Further information on the differential geometric aspects of the 2DTL may be found in Fordy and Wood [1994].

The primitive maps which occur in the special case of the finite open 2DTL are examples of *super-horizontal holomorphic* maps. The idea of passing from such maps to harmonic maps is older: It is an example of the *twistor construction*. The twistor construction is a fruitful method of constructing harmonic maps $\mathbf{C} \to G/K$, and it has played a prominent role in the development of the subject. Some examples appear in Eells and Lemaire [1988], and a general treatment is given in Burstall and Rawnsley [1990] (where the reader will find many other references). From the point of view of loop groups, the harmonic maps produced by the twistor construction are characterized by the property that the corresponding solutions of (Ω) are fixed (pointwise) by the S^1-action on ΩG. In the case $G = U_n$, the corresponding solutions of (Gr) are those of the form

$$W = \sum_{i \geq 0} A_i \lambda^i \begin{pmatrix} \lambda^{k_1} & & \\ & \ddots & \\ & & \lambda^{k_n} \end{pmatrix} H_+^{(n)}$$

where $\sum_{i \geq 0} A_i \lambda^i$ is constant in λ, i.e., $A_i = 0$ for $i \geq 1$. We saw in Chapter 20 that all harmonic maps $S^2 \to \mathbf{C}P^n$ have this property.

Chapter 22: Weierstrass formulae for harmonic maps

The method of Chapter 20 leads to *explicit formulae* for harmonic maps of finite uniton number. These are analogous to the explicit solutions of the 1DTL, in the sense that they are given by "factorizing an exponential". In this case, however, the initial data consists of a (matrix-valued) meromorphic function. The basis for these formulae is the Bruhat decomposition of the algebraic Grassmannian $\mathrm{Gr}^{(n),\mathrm{alg}}$.

I. Motivation and examples.

The Bruhat decomposition of $\mathrm{Gr}_k(\mathbf{C}^n)$ is a decomposition into algebraically embedded complex cells. Our formulae are based on the principle that, if $f : \mathbf{C} \to \mathrm{Gr}_k(\mathbf{C}^n)$ is a holomorphic map, then $f(\mathbf{C}-D)$ is contained in precisely one of these cells, for some discrete subset D. In Chapter 20 we made use of this principle to classify certain kinds of harmonic maps, and now we use it for a different purpose. We begin by considering two familiar examples, namely holomorphic maps $f : \mathbf{C} \to \mathrm{Gr}_k(\mathbf{C}^n)$ and harmonic maps $S^2 \to \mathbf{C}P^n$.

Let $f : \mathbf{C} \to \mathrm{Gr}_k(\mathbf{C}^n)$ be holomorphic. By the above principle, $f(\mathbf{C}-D)$ is contained in a Bruhat cell. Let us assume that this cell is the big cell, i.e., $N'_n \mathbf{C}^k$. (This is the generic situation, and in any case the other cells may be dealt with in a similar way.) Now, by the same algebraic observation as in Chapter 14 for $\mathbf{C}P^{n-1}$, we have

$$N'_n \mathbf{C}^k = \left\{ \begin{pmatrix} I_k & 0 \\ X & I_{n-k} \end{pmatrix} \mathbf{C}^k \ \middle| \ X \text{ is an } n-k \times k \text{ complex matrix} \right\}$$

$$\cong \mathbf{C}^{n-k} \times \mathbf{C}^k.$$

Hence we may write

$$f = \begin{pmatrix} I_k & 0 \\ c & I_{n-k} \end{pmatrix} \mathbf{C}^k = \exp \begin{pmatrix} 0 & 0 \\ c & 0 \end{pmatrix} \mathbf{C}^k$$

for some meromorphic $n - k \times k$ matrix-valued function c.

Let us re-phrase this in terms of (Gr). We have a harmonic map $\phi = i \circ f : \mathbf{C} \to U_n$, where $i : \mathrm{Gr}_k(\mathbf{C}^n) \to U_n$ is the usual inclusion $V \mapsto \pi_V - \pi_V^{\perp}$. Corresponding to ϕ we have (from Chapter 17) a solution of (Gr), given by

$$W = f \oplus \lambda H_+^{(n)}.$$

By the remarks above we can re-write this as

$$
\begin{aligned}
W &= \exp\begin{pmatrix} 0 & 0 \\ c & 0 \end{pmatrix} \mathbf{C}^k \oplus \lambda H_+^{(n)} \\
&= \exp\begin{pmatrix} 0 & 0 \\ c & 0 \end{pmatrix} (P_k + \lambda P_k^\perp) H_+^{(n)} \\
&= (P_k + \lambda P_k^\perp)(P_k + \tfrac{1}{\lambda} P_k^\perp)\exp\begin{pmatrix} 0 & 0 \\ c & 0 \end{pmatrix}(P_k + \lambda P_k^\perp) H_+^{(n)} \\
&= (P_k + \lambda P_k^\perp)\exp\begin{pmatrix} 0 & 0 \\ \lambda^{-1}c & 0 \end{pmatrix} H_+^{(n)}.
\end{aligned}
$$

(Here, P_k denotes projection on $\operatorname{Span}\{e_1,\ldots,e_k\}$.) Ignoring the constant loop $P_k + \lambda P_k^\perp$, we conclude that the corresponding solution of (Ω) is given by

$$
F = \left[\exp\begin{pmatrix} 0 & 0 \\ \lambda^{-1}c & 0 \end{pmatrix}\right]_u,
$$

where the notation $\gamma = \gamma_u \gamma_+$ denotes the factorization of a loop γ with respect to the Iwasawa decomposition $\Lambda GL_n \mathbf{C} = \Omega U_n \Lambda_+ GL_n \mathbf{C}$. This is our Weierstrass formula, for holomorphic maps $\mathbf{C} \to \operatorname{Gr}_k(\mathbf{C}^n)$.

Next, we consider harmonic maps $S^2 \to \mathbf{C}P^n$. By the classification theorem of Chapter 20, any such map is of the form

$$
f_i : \mathbf{C} \to \mathbf{C}P^n, \quad f_i = [\operatorname{Span}\{\tilde{f},\tilde{f}',\ldots,\tilde{f}^{(i-1)}\}^\perp \cap \operatorname{Span}\{\tilde{f},\tilde{f}',\ldots,\tilde{f}^{(i)}\}],
$$

and a corresponding solution of (Gr) is

$$
\begin{aligned}
W &= \pi_{i-1} \oplus \lambda \pi_i \oplus \lambda^2 H_+^{(n+1)} \\
&= \begin{pmatrix} | & | & & | \\ \tilde{f} & \tilde{f}' & \cdots & \tilde{f}^{(n)} \\ | & | & & | \end{pmatrix}(P_{i-1} + \lambda P_{i-1}^\perp P_i + \lambda^2 P_i^\perp) H_+^{(n+1)}.
\end{aligned}
$$

(In this formula, P_i denotes projection on $\operatorname{Span}\{e_0,\ldots,e_i\}$, and π_i denotes projection on $\operatorname{Span}\{\tilde{f},\tilde{f}',\ldots,\tilde{f}^{(i)}\}$.) A corresponding solution of (Ω) is given by the factorization formula

$$
F = \left[\begin{pmatrix} | & | & & | \\ \tilde{f} & \tilde{f}' & \cdots & \tilde{f}^{(n)} \\ | & | & & | \end{pmatrix}(P_{i-1} + \lambda P_{i-1}^\perp P_i + \lambda^2 P_i^\perp)\right]_u.
$$

As in Chapter 20, the factorization here amounts to performing the Gram-Schmidt orthogonalization on $\tilde{f},\tilde{f}',\ldots,\tilde{f}^{(n)}$.

The above solution W of (Gr) may be identified with the map (π_{i-1}, π_i) into the flag manifold $F_{i,i+1}(\mathbf{C}^{n+1}) = GL_{n+1}\mathbf{C}(\mathbf{C}^i \subseteq \mathbf{C}^{i+1} \subseteq \mathbf{C}^{n+1})$. It is then possible to write W in exponential form, by using the Bruhat decomposition of $F_{i,i+1}(\mathbf{C}^{n+1})$. We shall do this in the simplest non-trivial case, where $F_{i,i+1}(\mathbf{C}^{n+1}) = F_{1,2}(\mathbf{C}^3)$. Our harmonic map in this case is given by $f_1 = [\mathrm{Span}\{\tilde{f}\}^\perp \cap \mathrm{Span}\{\tilde{f}, \tilde{f}'\}]$. Since f_0 is holomorphic, and f_2 is anti-holomorphic, f_1 represents the most general harmonic map $S^2 \to \mathbf{C}P^2$ which is neither holomorphic nor anti-holomorphic.

As in the previous example, let us assume that the image of $W|_{\mathbf{C}-D}$ lies in the big cell N_3' ($\mathbf{C} \subseteq \mathbf{C}^2 \subseteq \mathbf{C}^3$), for some discrete subset D. Equivalently, we assume that we have a factorization

$$
\begin{pmatrix} | & | & | \\ \tilde{f} & \tilde{f}' & \tilde{f}'' \\ | & | & | \end{pmatrix} = \begin{pmatrix} 1 & 0 & 0 \\ \alpha & 1 & 0 \\ \beta & \gamma & 1 \end{pmatrix} \begin{pmatrix} * & * & * \\ 0 & * & * \\ 0 & 0 & * \end{pmatrix} \quad (\in N_3' \hat{N}_3).
$$

Since

$$
\begin{pmatrix} 1 & & \\ & \lambda & \\ & & \lambda^2 \end{pmatrix}^{-1} \hat{N}_3 \begin{pmatrix} 1 & & \\ & \lambda & \\ & & \lambda^2 \end{pmatrix} \subseteq \Lambda_+ GL_3\mathbf{C},
$$

we obtain

$$
W = \begin{pmatrix} 1 & 0 & 0 \\ \alpha & 1 & 0 \\ \beta & \gamma & 1 \end{pmatrix} \begin{pmatrix} 1 & & \\ & \lambda & \\ & & \lambda^2 \end{pmatrix} H_+^{(3)}
$$

$$
= \mathbf{C} \begin{pmatrix} 1 \\ \alpha \\ \beta \end{pmatrix} \oplus \lambda \mathbf{C} \begin{pmatrix} 1 \\ \alpha \\ \beta \end{pmatrix} \oplus \lambda \mathbf{C} \begin{pmatrix} 0 \\ 1 \\ \gamma \end{pmatrix} \oplus \lambda^2 H_+^{(3)}.
$$

It follows from the condition $\frac{\partial}{\partial z}\pi_0 \subseteq \pi_1$ that $\gamma = \beta'/\alpha'$, where $\alpha' = \partial\alpha/\partial z$ etc. Conversely, for any meromorphic functions α and β, the formula

$$
W = \begin{pmatrix} 1 & 0 & 0 \\ \alpha & 1 & 0 \\ \beta & \beta'/\alpha' & 1 \end{pmatrix} \begin{pmatrix} 1 & & \\ & \lambda & \\ & & \lambda^2 \end{pmatrix} H_+^{(3)}
$$

defines a solution of (Gr) which corresponds to a harmonic map $S^2 \to \mathbf{C}P^2$.

Now, we have

$$
\begin{pmatrix} 1 & 0 & 0 \\ \alpha & 1 & 0 \\ \beta & \beta'/\alpha' & 1 \end{pmatrix} = \exp \begin{pmatrix} 0 & 0 & 0 \\ a & 0 & 0 \\ b & c & 0 \end{pmatrix}
$$

where $a = \alpha$, $b = \beta - \frac{1}{2}\alpha(\alpha'/\beta')$, $c = \alpha'/\beta'$. As in the first example, we can re-write W as follows:

$$W = \begin{pmatrix} 1 & & \\ & \lambda & \\ & & \lambda^2 \end{pmatrix} \begin{pmatrix} 1 & & \\ & \lambda & \\ & & \lambda^2 \end{pmatrix}^{-1} \exp \begin{pmatrix} 0 & 0 & 0 \\ a & 0 & 0 \\ b & c & 0 \end{pmatrix} \begin{pmatrix} 1 & & \\ & \lambda & \\ & & \lambda^2 \end{pmatrix} H_+^{(3)}$$

$$= \begin{pmatrix} 1 & & \\ & \lambda & \\ & & \lambda^2 \end{pmatrix} \exp \begin{pmatrix} 0 & 0 & 0 \\ \lambda^{-1}a & 0 & 0 \\ \lambda^{-2}b & \lambda^{-1}c & 0 \end{pmatrix} H_+^{(3)}.$$

Ignoring the constant loop on the left, the corresponding solution of (Ω) is

$$F = \left[\exp \begin{pmatrix} 0 & 0 & 0 \\ \lambda^{-1}a & 0 & 0 \\ \lambda^{-2}b & \lambda^{-1}c & 0 \end{pmatrix} \right]_u .$$

This is our Weierstrass formula, for a harmonic map $S^2 \to \mathbb{C}P^2$ which is neither holomorphic nor anti-holomorphic. Note that the formula depends only on the choice of two meromorphic functions α and β.

II. Weierstrass formulae for harmonic maps $S^2 \to U_n$.

The special examples of the previous section involve no more than the Bruhat decomposition of a finite dimensional Grassmannian or flag manifold. To deal with more general harmonic maps, we need the Bruhat decomposition of $\mathrm{Gr}^{(n),\mathrm{alg}}$.

Let $\phi : \mathbb{C} \to U_n$ be a harmonic map of finite uniton number. For example, any harmonic map $S^2 \to U_n$ gives (on restriction to $\mathbb{C}$) such a map. As in section II of Chapter 20, we can write a solution of (Gr) corresponding to ϕ in the form

$$W = \sum_{i \geq 0} A_i \lambda^i \begin{pmatrix} \lambda^{k_1} & & \\ & \ddots & \\ & & \lambda^{k_n} \end{pmatrix} H_+^{(n)}$$

with[6]

$$k_1 = \cdots = k_{r_0} - k$$
$$k_{r_0+1} = \cdots = k_{r_1} = k - 1$$
$$\cdots \quad \cdots$$
$$k_{r_{s-1}+1} = \cdots = k_{r_s} = k_n = 1.$$

---

[6]There is a slight difference of notation between this section and examples of the previous section: In the previous section we have $k_1 \leq \cdots \leq k_n$, whereas in this section $k_1 \geq \cdots \geq k_n$. We have made no attempt to remedy this, because each convention appears naturally when first used. Moreover, it is useful to have both conventions available.

The n-tuple $(k_1, \ldots, k_n)$ may be regarded as the "type" of ϕ; this is a refinement of the concept of uniton number.

Let us write $\mathbf{C}^n = U_0 \oplus \cdots \oplus U_s$, where

$$U_0 = V_1 \oplus \cdots \oplus V_{r_0}, \quad U_1 = V_{r_0+1} \oplus \cdots \oplus V_{r_1}, \quad \ldots, \quad U_s = V_{r_{s-1}+1} \oplus \cdots \oplus V_{r_s}.$$

If X is an $n \times n$ matrix, we write $X = (X^{ab})$, where X^{ab} is the (a,b)-th "block" of X with respect to the decomposition $\mathbf{C}^n = U_0 \oplus \cdots \oplus U_s$ (with $0 \leq a, b \leq s$).

We claim that the expression for W can be simplified further:

Theorem.

$$W = \exp\left(\sum_{i=0}^{k-1} C_i \lambda^i\right) \begin{pmatrix} \lambda^{k_1} & & \\ & \ddots & \\ & & \lambda^{k_n} \end{pmatrix} H_+^{(n)},$$

where each C_i is a meromorphic $n \times n$ matrix-valued function, such that $C_i^{ab} = 0$ for $a \geq b - i$.

Sketch of the proof. What we need here is an explicit description of the Bruhat cell

$$\Lambda_+^{\mathrm{alg}} GL_n \mathbf{C} \begin{pmatrix} \lambda^{k_1} & & \\ & \ddots & \\ & & \lambda^{k_n} \end{pmatrix} H_+^{(n)}.$$

Such a description was given in Bott [1956], and more recent versions are given in Pressley [1982]; Burstall and Guest [preprint]. We just give a brief sketch here.

First, by an extension of the argument of section II of Chapter 17, W "collapses" to the form

$$W = \sum_{i=0}^{k-1} B_i \lambda^i \begin{pmatrix} \lambda^{k_1} & & \\ & \ddots & \\ & & \lambda^{k_n} \end{pmatrix} H_+^{(n)}.$$

(The theorem of that section is for the case $k = 1$.) Next, by considering the isotropy subgroup at

$$\begin{pmatrix} \lambda^{k_1} & & \\ & \ddots & \\ & & \lambda^{k_n} \end{pmatrix} H_+^{(n)}$$

for the action of the group $\Lambda_+^{\mathrm{alg}} GL_n \mathbf{C}$, it can be shown that it suffices to take $B_i = (B_i^{ab})$ such that $B_i^{ab} = 0$ if $a \geq b - i$. Finally, the stated formula follows from this because of the fact that the exponential map is a bi-holomorphism for any nilpotent (complex) Lie group. The functions C_i are meromorphic, because W is holomorphic (off a discrete set). ∎

We may re-write the above expression for W as

$$
\begin{pmatrix} \lambda^{k_1} & & \\ & \ddots & \\ & & \lambda^{k_n} \end{pmatrix}^{-1} W =
$$

$$
\exp\left[\begin{pmatrix} \lambda^{k_1} & & \\ & \ddots & \\ & & \lambda^{k_n} \end{pmatrix}^{-1} \sum_{i=0}^{k-1} C_i \lambda^i \begin{pmatrix} \lambda^{k_1} & & \\ & \ddots & \\ & & \lambda^{k_n} \end{pmatrix} \right] H_+^{(n)}.
$$

Ignoring the first (constant) factor, and converting back from (Gr) to (Ω), we obtain the following "factorization of exponentials" formula for solutions of (Ω):

Theorem (Weierstrass formula). *Let $F : S^2 \to \Omega^{\mathrm{alg}} U_n$ be a solution of* (Ω) *which is non-degenerate in the sense of Chapter 20. Then F may be written in the form*

$$
F = \left[\exp \begin{pmatrix} \lambda^{k_1} & & \\ & \ddots & \\ & & \lambda^{k_n} \end{pmatrix}^{-1} \sum_{i=0}^{k-1} C_i \lambda^i \begin{pmatrix} \lambda^{k_1} & & \\ & \ddots & \\ & & \lambda^{k_n} \end{pmatrix} \right]_u ,
$$

where $k_1, \ldots, k_n$ and $C_0, \ldots, C_{k-1}$ are as above. ∎

We do not claim that this formula gives a solution of (Ω) for *arbitrary* meromorphic functions C_i, however. The C_i are required to satisfy certain differential equations, corresponding to the condition $\frac{\partial}{\partial z} W \subseteq \lambda^{-1} W$. These may be solved fairly explicitly "by integration", and are described in Burstall and Guest [preprint]. We just illustrate the situation for a specific example.

Example: Harmonic maps $S^2 \to U_3$

There are three possible types in this case, corresponding to $(k_1, k_2, k_3) = (2, 1, 0)$, $(1, 1, 0)$, or $(1, 0, 0)$. (We ignore the case $(0, 0, 0)$, which gives constant W.) The expressions for W take the following forms:

Type 1: $\quad W = (B_0 + \lambda B_1) \begin{pmatrix} \lambda^2 & & \\ & \lambda & \\ & & 1 \end{pmatrix} H_+^{(3)}.$

Type 2: $\quad W = B_0 \begin{pmatrix} \lambda & & \\ & \lambda & \\ & & 1 \end{pmatrix} H_+^{(3)} = B_0 V_3 \oplus \lambda H_+^{(3)}.$

Type 3:　$W = B_0 \begin{pmatrix} \lambda & & \\ & 1 & \\ & & 1 \end{pmatrix}$　$H_+^{(3)} = B_0(V_3 \oplus V_2) \oplus \lambda H_+^{(3)}.$

Types 2 and 3 correspond to holomorphic maps into $\mathrm{Gr}_1(\mathbf{C}^3) = \mathbf{C}P^2$ and $\mathrm{Gr}_2(\mathbf{C}^3)$ $(\cong \mathbf{C}P^2)$, and may be dealt with as in section I.

Type 1 is the most interesting case. By the first theorem above, W is actually of the form

$$W = \exp\left[\begin{pmatrix} 0 & a & b \\ 0 & 0 & c \\ 0 & 0 & 0 \end{pmatrix} + \lambda \begin{pmatrix} 0 & 0 & d \\ 0 & 0 & 0 \\ 0 & 0 & 0 \end{pmatrix}\right] \begin{pmatrix} \lambda^2 & & \\ & \lambda & \\ & & 1 \end{pmatrix} H_+^{(3)}$$

for some meromorphic functions a, b, c, d. The differential equation corresponding to the condition $\frac{\partial}{\partial z} W \subseteq \lambda^{-1} W$ may be found by inspection here; it is

$$2b' = ac' - a'c$$

(where $a' = \partial a/\partial z$, etc.). In particular, it turns out that there is no condition on d. The solutions to this equation are given by all pairs (a, c) of meromorphic (i.e., rational) functions such that $ac' - a'c$ has no residues. Alternatively, motivated by section I, we may make the change of variables

$$\alpha = b + \tfrac{1}{2}ac, \quad \beta = c, \quad \gamma = a.$$

The differential equation simplifies to $\alpha'/\beta' = \gamma$, and so its solutions are given by arbitrary pairs (α, β) of meromorphic functions.

We have

$$\begin{pmatrix} \lambda^2 & & \\ & \lambda & \\ & & 1 \end{pmatrix}^{-1} W =$$

$$\exp\left[\begin{pmatrix} \lambda^2 & & \\ & \lambda & \\ & & 1 \end{pmatrix}^{-1} \begin{pmatrix} 0 & a & b + \lambda d \\ 0 & 0 & c \\ 0 & 0 & 0 \end{pmatrix} \begin{pmatrix} \lambda^2 & & \\ & \lambda & \\ & & 1 \end{pmatrix}\right] H_+^{(3)}.$$

The corresponding solution of (Ω) is

$$F = \left[\exp \begin{pmatrix} 0 & \lambda^{-1}a & \lambda^{-2}b + \lambda^{-1}d \\ 0 & 0 & \lambda^{-1}c \\ 0 & 0 & 0 \end{pmatrix}\right]_u.$$

This is the Weierstrass formula for a (non-degenerate) harmonic map $S^2 \to U_3$ of type 1. It depends only on the three meromorphic functions α, β, d. In the special case where d is identically zero, we recover the Weierstrass formula of section I for harmonic maps $S^2 \to \mathbf{C}P^2$.

III. Weierstrass formulae for harmonic maps $S^2 \to \mathrm{Gr}_k(\mathbf{C}^n)$.

We can obtain Weierstrass formulae for harmonic maps $\phi : \mathbf{C} \to \mathrm{Gr}_k(\mathbf{C}^n)$ of finite uniton number in a similar way. From section II of Chapter 20, we know such maps correspond to solutions $\{W_i\} : \mathbf{C} \to \mathrm{Fl}_{k,n-k}^{(n)}$ of (Fl) of the form

$$\{W_i\} = \sum_{i \geq 0} A_i \lambda^i \begin{pmatrix} \lambda^{k_1} & & \\ & \ddots & \\ & & \lambda^{k_n} \end{pmatrix} \qquad (\lambda H_+^{(n)} \subseteq V_{i_1} \oplus \cdots \oplus V_{i_k} \oplus \lambda H_+^{(n)} \subseteq H_+^{(n)}).$$

The conditions on $k_1, \ldots, k_n$ here are:

$$k_1 = \cdots = k_{r_0} = l$$
$$k_{r_0+1} = \cdots = k_{r_1} = l - 1$$
$$\cdots \qquad \cdots$$
$$k_{r_{s-1}+1} = \cdots = k_{r_s} = k_n = 1.$$

The conditions on $i_1, \ldots, i_k$ are

$$\{r_{s-2} + 1, \ldots, r_{s-1}\} \neq S_1 \neq \emptyset$$
$$\{r_{s-3} + 1, \ldots, r_{s-2}\} \neq S_2 \neq \emptyset$$
$$\cdots \qquad \cdots$$
$$\{r_0 + 1, \ldots, r_1\} \neq S_{s-1} \neq \emptyset$$
$$S_s \neq \emptyset$$

(where $S_t = \{j \mid r_{s-t-1} + 1 \leq i_j \leq r_{s-t}\}$, as in Chapter 20).

These conditions are rather restrictive. For $\mathrm{Gr}_2(\mathbf{C}^4)$ the only possibilities are

Type 1: $\{W_i\} =$

$$(B_0 + \lambda B_1) \begin{pmatrix} \lambda^2 & & & \\ & \lambda & & \\ & & \lambda & \\ & & & 1 \end{pmatrix} \qquad (\lambda H_+^{(4)} \subseteq V_i \oplus V_j \oplus \lambda H_+^{(4)} \subseteq H_+^{(4)}),$$

where $(i, j) = (1, 2)$ or $(1, 3)$.

Type 2: $\{W_i\} = B_0 \begin{pmatrix} \lambda & & & \\ & \lambda & & \\ & & \lambda & \\ & & & 1 \end{pmatrix} \qquad (\lambda H_+^{(4)} \subseteq V_i \oplus V_j \oplus \lambda H_+^{(4)} \subseteq H_+^{(4)}),$

where $(i, j) = (1, 2)$, $(1, 3)$, or $(2, 3)$.

Type 3: $\{W_i\} = B_0 \begin{pmatrix} \lambda & & & \\ & \lambda & & \\ & & 1 & \\ & & & 1 \end{pmatrix} \qquad (\lambda H_+^{(4)} \subseteq V_1 \oplus V_2 \oplus \lambda H_+^{(4)} \subseteq H_+^{(4)}).$

To obtain corresponding solutions of (Gr), we apply the appropriate isomorphism $T : \Lambda GL_4 \mathbf{C} \to (\Lambda GL_4 \mathbf{C})_\sigma$. For example, for $\{W_i\}$ of type 3, we have

$$W = T\{W_i\}$$

$$= \begin{pmatrix} \lambda & & & \\ & \lambda & & \\ & & 1 & \\ & & & 1 \end{pmatrix} B_0 \begin{pmatrix} \lambda^2 & & & \\ & \lambda^2 & & \\ & & 1 & \\ & & & 1 \end{pmatrix} \begin{pmatrix} \lambda^{-1} & & & \\ & \lambda^{-1} & & \\ & & 1 & \\ & & & 1 \end{pmatrix} H_+^{(4)}$$

$$= \begin{pmatrix} \lambda & & & \\ & \lambda & & \\ & & 1 & \\ & & & 1 \end{pmatrix} B_0 \begin{pmatrix} \lambda & & & \\ & \lambda & & \\ & & 1 & \\ & & & 1 \end{pmatrix} H_+^{(4)}$$

$$= \begin{pmatrix} \lambda^2 & & & \\ & \lambda^2 & & \\ & & 1 & \\ & & & 1 \end{pmatrix} \exp \begin{pmatrix} 0 & \lambda^{-1}c \\ 0 & 0 \end{pmatrix} H_+^{(4)}.$$

This agrees (up to minor changes of notation) with the formula in section I for holomorphic maps $\mathbf{C} \to \mathrm{Gr}_2(\mathbf{C}^4)$. We conclude that the harmonic maps of type 3 are precisely the holomorphic maps.

Maps of type 2 are of the form $\phi = f^\perp \cap g$, where (f, g) is a holomorphic map $S^2 \to \mathrm{F}_{1,3}(\mathbf{C}^4)$ with $\frac{\partial}{\partial z} f \subseteq g$. These have uniton number 2.

Maps of type 1 are the most complicated. They have uniton number 3. Taking $(i, j) = (1, 2)$, the corresponding solutions of (Gr) are given by Weierstrass formulae of the form

$$W = \exp \left[\begin{pmatrix} 0 & a & b & c \\ 0 & 0 & d & e \\ 0 & 0 & 0 & f \\ 0 & 0 & 0 & 0 \end{pmatrix} + \lambda^2 \begin{pmatrix} 0 & 0 & 0 & g \\ 0 & 0 & 0 & 0 \\ 0 & 0 & 0 & 0 \\ 0 & 0 & 0 & 0 \end{pmatrix} \right] \begin{pmatrix} \lambda^3 & & & \\ & \lambda^2 & & \\ & & \lambda & \\ & & & 1 \end{pmatrix} H_+^{(4)}.$$

Here, a, b, c, d, e, f are determined by

$$\exp \begin{pmatrix} 0 & a & b & c \\ 0 & 0 & d & e \\ 0 & 0 & 0 & f \\ 0 & 0 & 0 & 0 \end{pmatrix} = \begin{pmatrix} 1 & \delta & \alpha'/\gamma' & \alpha \\ 0 & 1 & \beta'/\gamma' & \beta \\ 0 & 0 & 1 & \gamma \\ 0 & 0 & 0 & 1 \end{pmatrix}, \quad \delta = (\alpha'/\gamma')'/(\beta'/\gamma')'$$

where α, β, γ, g are arbitrary rational functions.

Bibliographical comments.

The term "Weierstrass formula" is used here because of the classical formula of Weierstrass for minimal surfaces in terms of holomorphic functions (see, for example, Berger and Gostiaux [1988]).

The fact that the dressing action on harmonic maps of finite uniton number "collapses" to the action of a finite dimensional group has been noted by various authors (Uhlenbeck [1989]; Arsenault and Saint-Aubin [1989a; 1989b]; Guest and Ohnita [1993]). The Bruhat decomposition of $\mathrm{Gr}^{(n),\mathrm{alg}}$ provides a simple explanation for this: The orbits of $\Lambda_+^{\mathrm{alg}} GL_n \mathbf{C}$ on $\mathrm{Gr}^{(n),\mathrm{alg}}$ are finite dimensional.

In Burstall and Guest [preprint] a slightly different point of view is taken, using ΩG itself (instead of the Grassmannian model of ΩG). It turns out to be remarkably easy to generalize the results of this chapter (and Chapter 20) to harmonic maps $S^2 \to G$ or G/K.

The Weierstrass formulae of this chapter provide examples of an observation of Dorfmeister *et al.* [in press], that *any* solution F of (Ω) can be written in the form $F = \Psi_u$, where $\Psi : \mathbf{C} - D \to \Lambda_- GL_n \mathbf{C}$ is holomorphic. From our point of view, this is a consequence of the Birkhoff decomposition. The relationship between this Ψ and our formulae is explained in Burstall and Guest [preprint].

Although our Weierstrass formulae (and the more general formulae of Burstall and Guest [preprint]) appear to be new, various related results have been known for some time. The special case of holomorphic maps $S^2 \to S^2$ is treated in Ward [1990]. A description of harmonic maps $S^2 \to U_3$, equivalent to our Weierstrass formula, was given in Wood [1989]; see also Piette and Zakrzewski [1988]. A description of harmonic maps $S^2 \to \mathrm{Gr}_2(\mathbf{C}^4)$ was given in Ramanathan [1984]; see also Sakagawa [in press]. Various generalizations to specific higher dimensional symmetric spaces are known; see Eells and Lemaire [1988]. Our formulae are somewhat simpler, perhaps, and easier to generalize. However, the main feature of our Weierstrass formulae is that they are all given by the fundamental procedure of "factorizing an exponential".

The idea of using Bruhat decompositions to obtain Weierstrass formulae was used in Bryant [1985], in the case of harmonic maps obtained via the twistor construction (cf. the bibliographical comments for Chapter 21). Our approach was motivated by Morse theory (see the bibliographical comments for Chapter 20), but it is equivalent to the approach of Bryant.

Part III

One-dimensional and two-dimensional integrable systems

Chapter 23: From 2 Lax equations to 1 zero-curvature equation

I. 1+1=2.

In Chapter 8 we saw that the solution $X : \mathbf{R} \to \mathbf{g}$ of the Lax equation

$$\dot{X} = [X, \pi_1 J_X], \quad X(0) = V$$

is given by the explicit formula

$$X(t) = \mathrm{Ad}(\exp tJ_V)_1^{-1} V.$$

Here, $J : \mathbf{g} \to \mathbf{g}$ is an "invariant vector field" on the Lie algebra $\mathbf{g}$ of G. (A typical example of such a J is given by the gradient $J = \nabla f$ of an invariant function $f : \mathbf{g} \to \mathbf{R}$.) We assume as usual that we have decompositions

$$G = G_1 G_2, \quad \mathbf{g} = \mathbf{g}_1 \oplus \mathbf{g}_2,$$

with respect to which we write

$$g = g_1 g_2, \quad W = \pi_1 W + \pi_2 W$$

for $g \in G, W \in \mathbf{g}$.

From section I of Chapter 8, recall that an invariant vector field J satisfies the property

(a) $[X, J_Y] = ([X, Y] \cdot J)_Y$, for all $X, Y \in \mathbf{g}$,

and that if J_1, J_2 are two invariant vector fields then

(b) $[(J_1)_X, (J_2)_X] = 0$ for all $X \in \mathbf{g}$.

Let us now consider *two* Lax equations of the above form:

(1) $$\frac{d}{ds} X_1 = [X_1, \pi_1(J_1)_{X_1}], \quad X_1(0) = V_1$$

(2) $$\frac{d}{dt} X_2 = [X_2, \pi_1(J_2)_{X_2}], \quad X_2(0) = V_2.$$

These are simply the equations for the flows defined by the vector fields $U, V : \mathbf{g} \to \mathbf{g}$, where

$$U_X = [X, \pi_1(J_1)_X]$$
$$V_X = [X, \pi_1(J_2)_X].$$

We show in this section that these flows commute, and hence give rise to a function $X(s, t)$. In the next section we show how this $X(s, t)$ gives rise to a solution of a zero-curvature equation.

The basis for these results is the following technical lemma:

Lemma. *With the above definitions, we have*

$$V \cdot \pi_1 J_1 - U \cdot \pi_1 J_2 = [\pi_1 J_1, \pi_1 J_2].$$

Proof. First, note that $V_X \cdot J_1 = [X, (\pi_1 J_2)_X] \cdot J_1 = -[(\pi_1 J_2)_X, J_1]$, by (a) above. So we have

$$
\begin{aligned}
V \cdot \pi_1 J_1 - U \cdot \pi_1 J_2 &= \pi_1([\pi_1 J_1, J_2] - [\pi_1 J_2, J_1]) \\
&= 2[\pi_1 J_1, \pi_1 J_2] + \pi_1([\pi_1 J_1, \pi_2 J_2] + [\pi_2 J_1, \pi_1 J_2])
\end{aligned}
$$

(by writing $J_i = \pi_1 J_i + \pi_2 J_i$). Since $[J_1, J_2] = 0$, by (b) above, we obtain

$$[\pi_1 J_1, \pi_2 J_2] + [\pi_2 J_1, \pi_1 J_2] + [\pi_1 J_1, \pi_1 J_2] + [\pi_2 J_1, \pi_2 J_2] = 0$$

(again by writing $J_i = \pi_1 J_i + \pi_2 J_i$). We conclude that

$$V \cdot \pi_1 J_1 - U \cdot \pi_1 J_2 = 2[\pi_1 J_1, \pi_1 J_2] + \pi_1(-[\pi_1 J_1, \pi_1 J_2]) = [\pi_1 J_1, \pi_1 J_2]$$

as required. ■

We use this to establish:

Proposition. *The Lie bracket of the vector fields* U, V *is zero.*

Proof. The Lie bracket of the vector fields U, V is $U \cdot V - V \cdot U$. (We avoid the usual notation $[U, V]$ for this Lie bracket, to avoid confusion with the pointwise Lie bracket of the **g**-valued functions U, V.) To compute this, note that

$$
\begin{aligned}
U \cdot V &= U \cdot [X, \pi_1 J_2] \\
&= [U \cdot X, \pi_1 J_2] + [X, U \cdot \pi_1 J_2] \\
&= [U, \pi_1 J_2] + [X, U \cdot \pi_1 J_2)] \\
&= [[X, \pi_1 J_1], \pi_1 J_2] + [X, U \cdot \pi_1 J_2)].
\end{aligned}
$$

So the desired Lie bracket is

$$
\begin{aligned}
U \cdot V - V \cdot U &= [[X, \pi_1 J_1], \pi_1 J_2] - [[X, \pi_1 J_2], \pi_1 J_1] + [X, U \cdot \pi_1 J_2 - V \cdot \pi_1 J_1] \\
&= [X, [\pi_1 J_1, \pi_1 J_2]] + [X, U \cdot \pi_1 J_2 - V \cdot \pi_1 J_1] \\
&\quad \text{(using the Jacobi identity)} \\
&= [X, U \cdot \pi_1 J_2 - V \cdot \pi_1 J_1 + [\pi_1 J_1, \pi_1 J_2]]
\end{aligned}
$$

which is zero by the lemma. ■

It follows that the individual flows of the vector fields U, V may be combined to give a "two parameter flow". This fact may be expressed more precisely as follows (see Chapter 5 of Spivak [1979]). Let $X_1^{V_1}$ be the solution of equation (1), and let $X_2^{V_2}$ be the solution of equation (2). Then there exists a neighbourhood N of $(0,0)$ in $\mathbf{R}^2$ and a function $X : N \to \mathbf{g}$ such that

$$X(s,t) = X_1^{V_1}(s) \quad \text{where } V_1 = X_2^V(t)$$
$$= X_2^{V_2}(t) \quad \text{where } V_2 = X_1^V(s).$$

This function X satisfies the system of equations

$$X_s = [X, \pi_1(J_1)x]$$
$$X_t = [X, \pi_1(J_2)x].$$

By using our explicit formulae for $X_1(s)$ and $X_2(t)$ we can obtain an explicit formula for $X(s,t)$:

Proposition. *The solution of the above system with* $X(0,0) = V$ *is given by* $X(s,t) = \mathrm{Ad}(\exp s(J_1)_V + t(J_2)_V)_1^{-1} V$.

Proof. By definition, putting $W = \mathrm{Ad}(\exp s(J_1)_V)_1^{-1} V$, we have

$$X(s,t) = \mathrm{Ad}(\exp t(J_2)_W)_1^{-1} W$$
$$= \mathrm{Ad}((\exp s(J_1)_V)_1(\exp t(J_2)_W)_1))^{-1} V.$$

Computing further, we obtain

$$(\exp s(J_1)_V)_1(\exp t(J_2)_W)_1$$
$$= (\exp s(J_1)_V)_1(\exp t\,\mathrm{Ad}(\exp s(J_1)_V)_1^{-1}(J_2)_V)_1$$
$$= (\exp s(J_1)_V)_1((\exp s(J_1)_V)_1^{-1}(\exp t(J_2)_V)(\exp s(J_1)_V)_1)_1$$
$$= (\exp t(J_2)_V(\exp s(J_1)_V)_1)_1$$
$$= (\exp t(J_2)_V \exp s(J_1)_V)_1$$
$$= (\exp t(J_2)_V + s(J_1)_V)_1 \text{ since } [(J_1)_V, (J_2)_V] = 0$$

as required. ∎

In particular, X is defined on the whole of $\mathbf{R}^2$.

II. A zero-curvature equation.

The "two-dimensional Lax equation"

$$X_s = [X, \pi_1(J_1)x]$$
$$X_t = [X, \pi_1(J_2)x]$$

is closely related to a zero-curvature equation:

Proposition. *Let $X : \mathbf{R}^2 \to \mathbf{g}$ be the solution of the above system, with $X(0,0) = V$. Then the functions $A = (\pi_1 J_1) \circ X$, $B = (\pi_1 J_2) \circ X$ satisfy the zero-curvature equation $A_t - B_s = [A, B]$.*

Proof. We have $A_t = (V \cdot \pi_1 J_1)_X$, and $B_s = (U \cdot \pi_1 J_2)_X$, where U, V are as in section I. Hence the proposition follows immediately from the lemma of that section. ∎

Because of the usual geometrical interpretation of a zero-curvature equation, there exists a unique function $F : \mathbf{R}^2 \to G_1$ such that

$$F^{-1}F_s = \pi_1(J_1) \circ X$$
$$F^{-1}F_t = \pi_1(J_2) \circ X$$

and $F(0,0) = e$. The relationship between F and X is given by:

Proposition. $X = \operatorname{Ad} F^{-1} V.$

Proof. It suffices to show that $Y = F^{-1}VF$ satisfies the same Lax equations as X. We have $Y_s = -F^{-1}F_s F^{-1}VF + F^{-1}VF_s = [Y, F^{-1}F_s]$, and similarly $Y_t = [Y, F^{-1}F_t]$. ∎

We can derive an explicit formula

$$F(s,t) = (\exp s(J_1)_V + t(J_2)_V)_1$$

for F, by the following argument. It suffices to show that F satisfies the equations $F^{-1}F_s = \pi_1(J_1) \circ X$, $F^{-1}F_t = \pi_1(J_2) \circ X$. Let $g(s,t) = \exp s(J_1)_V + t(J_2)_V$. Then

$$
\begin{aligned}
(J_1)_V &= (\exp s(J_1)_V + t(J_2)_V)_s (\exp s(J_1)_V + t(J_2)_V)^{-1} \\
&= (g_1 g_2)_s (g_1 g_2)^{-1} \\
&= (g_1)_s g_1^{-1} + g_1 (g_2)_s g_2^{-1} g_1^{-1}
\end{aligned}
$$

and so $g_1^{-1}(J_1)_V g_1 = g_1^{-1}(g_1)_s + (g_2)_s g_2^{-1}$. Hence $F^{-1}F_s = g_1^{-1}(g_1)_s = \pi_1 g_1^{-1}(J_1)_V g_1 = (\pi_1 J_1)_X$. Similarly we have $F^{-1}F_t = (\pi_1 J_2)_X$. (This argument is the one used in Chapter 5 to verify the explicit solution of a one-dimensional Lax equation.)

We therefore obtain the explicit formula

$$X(s,t) = \operatorname{Ad}(\exp s(J_1)_V + t(J_2)_V)_1^{-1} V$$

for X, in agreement with the formula from section I.

Bibliographical comments.

The "$1 + 1 = 2$" principle is based on Theorem 2.1 of Ferus *et al.* [1992].

Chapter 24: Harmonic maps of finite type

I. Harmonic maps of finite type.

The construction of Chapter 23 gives rise to harmonic maps, if we make appropriate choices for (1) the group decomposition, and (2) the vector fields J_1, J_2.

For the group decomposition we choose the Iwasawa decomposition of a loop group. More specifically, we choose

$$\Lambda G^c = \Omega G \, \Lambda_+ G^c$$

where G is a compact Lie group. Any $\gamma \in \Lambda G^c$ has a factorization $\gamma = \gamma_u \gamma_+$, and any $f \in \Lambda \mathbf{g} \otimes \mathbf{C}$ may be written $f = \pi_u f + \pi_+ f$, where $\pi_u f \in \Omega \mathbf{g}$, $\pi_+ f \in \Lambda_+ \mathbf{g} \otimes \mathbf{C}$.

For the vector fields $J_1, J_2 : \Lambda \mathbf{g} \otimes \mathbf{C} \to \Lambda \mathbf{g} \otimes \mathbf{C}$ we choose

$$(J_1)_f = \lambda^{d-1} f$$
$$(J_2)_f = \sqrt{-1}\, \lambda^{d-1} f$$

where d is a fixed non-negative integer.

In contrast to the situation of Chapter 23, the Lie groups and Lie algebras are infinite dimensional here. We consider two independent Lax equations

$$\frac{d}{ds} X_1 = [X_1, \pi_u(J_1)_{X_1}]$$
$$\frac{d}{dt} X_2 = [X_2, \pi_u(J_2)_{X_2}],$$

with initial value $V(\lambda) = \sum_{i=-d}^{d} V_i \lambda^i$. As in Chapter 23, the solutions X_1 and X_2 may be combined into a single solution X of the system

$$X_s = [X, \pi_u(J_1)_X]$$
$$X_t = [X, \pi_u(J_2)_X],$$

and we have the explicit formula $X = \mathrm{Ad}\, F^{-1} V$, where

$$F(s,t) = (\exp s(J_1)_V + t(J_2)_V)_u.$$

If the solution to each Lax equation (and hence also X) remains within the finite dimensional subspace

$$\{f \in \Lambda \mathbf{g} \otimes \mathbf{C} \mid f(\lambda) = \sum_{i=-d}^{d} A_i \lambda^i\},$$

then we say that V is an initial condition of *finite type*.

Let us turn our attention to the map F. As in Chapter 23, F is a solution of the system

$$F^{-1}F_s = \pi_u(J_1)$$
$$F^{-1}F_t = \pi_u(J_2).$$

We claim that F is actually a solution to the *harmonic map equation* (Ω):

Proposition. *Assume that V is an initial condition of finite type. Then, writing $z = s + \sqrt{-1}\,t$, we have*

$$F^{-1}F_z = \text{linear in } \tfrac{1}{\lambda}$$
$$F^{-1}F_{\bar{z}} = \text{linear in } \lambda.$$

Hence, F is a solution of (Ω).

Proof. We need an explicit description of the projection map $\pi_u : \Lambda\mathbf{g}\otimes\mathbf{C} \to \Omega\mathbf{g}$, i.e., the projection map onto the first summand of

$$\Lambda\mathbf{g} \otimes \mathbf{C} = (\Omega\mathbf{g}) \oplus (\Lambda_+\mathbf{g} \otimes \mathbf{C}).$$

Let us write the Fourier series of a loop $f \in \Lambda\mathbf{g} \otimes \mathbf{C}$ in the form $f(\lambda) = A_0 + \sum_{i \neq 0}(1 - \lambda^i)A_i$. We write $f_+(\lambda) = \sum_{i>0}(1 - \lambda^i)A_i$, and $f_-(\lambda) = \sum_{i<0}(1 - \lambda^i)A_i$. Let $c : \mathbf{g} \otimes \mathbf{C} \to \mathbf{g} \otimes \mathbf{C}$ be the involution with respect to the real form $\mathbf{g}$. The corresponding involution of $\Lambda\mathbf{g} \otimes \mathbf{C}$, also denoted c, is determined by $c(A\lambda^i) = c(A)\lambda^{-i}$.

From the identity

$$f = (f_- + c(f_-)) + (f_+ - c(f_-) + A_0)$$

we see that $\pi_u(f) = f_- + c(f_-)$. For $X = X_0 + \sum_{i=-d}^{d} X_i(1 - \lambda^i)$, we have

$$\pi_u(J_1)_X = \pi_u(\lambda^{d-1}X) = (1 - \tfrac{1}{\lambda})X_{-d} + c((1 - \tfrac{1}{\lambda})X_{-d})$$
$$\pi_u(J_2)_X = \pi_u(\sqrt{-1}\,\lambda^{d-1}X) = \sqrt{-1}\,(1 - \tfrac{1}{\lambda})X_{-d} + c(\sqrt{-1}\,(1 - \tfrac{1}{\lambda})X_{-d}).$$

Conversion to complex derivatives gives:

$$F^{-1}F_z = \tfrac{1}{2}(F^{-1}F_s - \sqrt{-1}\,F^{-1}F_t) = \tfrac{1}{2}\pi_u(J_1 - \sqrt{-1}\,J_2)$$
$$F^{-1}F_{\bar{z}} = \tfrac{1}{2}(F^{-1}F_s + \sqrt{-1}\,F^{-1}F_t) = \tfrac{1}{2}\pi_u(J_1 + \sqrt{-1}\,J_2).$$

Hence

$$F^{-1}F_z = (1 - \tfrac{1}{\lambda})X_{-d}$$
$$F^{-1}F_{\bar{z}} = (1 - \lambda)c(X_{-d}),$$

as required. ■

Note that $\phi^{-1}\phi_z = F^{-1}F_z|_{\lambda=-1} = 2X_{-d}$.

From the proof of the proposition, we can express more explicitly the ordinary differential equations which are responsible for our harmonic maps. These equations are

$$X_z = [X, (1 - \tfrac{1}{\lambda})X_{-d}]$$
$$X_{\bar{z}} = [X, (1 - \lambda)c(X_{-d})].$$

Following Pinkall and Sterling [1989]; Ferus *et al.* [1992]; Burstall *et al.* [1993], we say that a harmonic map $\phi(z) = F(z, -1)$ constructed from such an X is a harmonic map of *finite type* (and, more precisely, of *type d*). We note the following formula for the corresponding solution of (Ω):

$$F(z, \lambda) = (\exp s(J_1)_V + t(J_2)_V)_u$$
$$= (\exp s(\lambda^{d-1}V) + t(\sqrt{-1}\,\lambda^{d-1}V))_u$$
$$= (\exp z\lambda^{d-1}V(\lambda))_u.$$

This is very similar to the Weierstrass formulae of Chapter 22!

The construction of X (and, therefore, of ϕ) depends only on the choice of the initial condition $V(\lambda) = \sum_{i=-d}^{d} V_i\lambda^i$. We may characterize such maps ϕ either in terms of X or in terms of F. To conclude this section, we summarize both possibilities:

(HF1)   Let $\phi : \mathbf{C} \to G$ be a harmonic map. Let d be a non-negative integer. Then ϕ has type d if and only if there exists some $V(\lambda) = \sum_{i=-d}^{d} V_i\lambda^i \in \Lambda\mathbf{g} \otimes \mathbf{C}$ such that $X_{-d} = \tfrac{1}{2}\phi^{-1}\phi_z$, where

$$X = X_0 + \sum_{i=-d}^{d} X_i(1 - \lambda^i) : \mathbf{C} \to \Lambda\mathbf{g} \otimes \mathbf{C}$$

is the solution of

$$X_z = [X, (1 - \tfrac{1}{\lambda})X_{-d}]$$
$$X_{\bar{z}} = [X, (1 - \lambda)c(X_{-d})]$$

with $X(0) = V$.

(HF2)   Let $\phi : \mathbf{C} \to G$ be a harmonic map. Let d be a non-negative integer. Then ϕ has type d if and only if there exists some $V(\lambda) = \sum_{i=-d}^{d} V_i\lambda^i \in \Lambda\mathbf{g} \otimes \mathbf{C}$ such that $F(z, \lambda) = (\exp z\lambda^{d-1}V(\lambda))_u$ is a solution of (Ω) corresponding to ϕ, and $X = F^{-1}VF$ is of the form $X = X_0 + \sum_{i=-d}^{d} X_i(1 - \lambda^i)$.

The condition that V be of finite type is implicit in these definitions. If V is real, however, it turns out that this condition is automatically satisfied, so *any* real V gives rise to a harmonic map of finite type:

Proposition. *If $V(\lambda) = \sum_{i=-d}^{d} V_i \lambda^i$ satisfies the reality condition $c(V) = V$, then V is an initial condition of finite type.*

Proof. We have to show that $X = F^{-1}VF$ takes values in the space

$$\{f \in \Lambda \mathbf{g} \otimes \mathbf{C} \mid f(\lambda) = \sum_{i=-d}^{d} A_i \lambda^i\}.$$

Let $G = (\exp z\lambda^{d-1}V)_+$, so that $\exp z\lambda^{d-1}V = FG$. Then $X = F^{-1}VF = GVG^{-1}$, so X contains no power of λ lower than λ^{-d}. By the reality condition, it follows that X contains no power of λ higher than λ^d. ∎

An alternative proof of this proposition may be given by showing directly that the vector field of the Lax equation is tangent to the above finite dimensional space. This is done in Pinkall and Sterling [1989]; Ferus *et al.* [1992]; Burstall *et al.* [1993], where V is *always* taken to be real.

II. Examples.

(1) The simplest examples are those of type 0. In this case, $X(z, \lambda) = X_0(z)$, and

$$(X_0)_z = [X_0, (1 - \tfrac{1}{\lambda})X_0]$$
$$(X_0)_{\bar{z}} = [X_0, (1 - \lambda)c(X_0)].$$

Comparing coefficients of λ, we find that $(X_0)_z = (X_0)_{\bar{z}} = 0$ (so X_0 is constant) and $[X_0, c(X_0)] = 0$. Let us write $X_0(z) = V \in \mathbf{g} \otimes \mathbf{C}$. Since V and $c(V)$ commute, we have the explicit factorization

$$\exp z\tfrac{1}{\lambda}V = (\exp z(\tfrac{1}{\lambda} - 1)V + \bar{z}(\lambda - 1)c(V))(\exp zV + \bar{z}c(V) - \bar{z}\lambda c(V)).$$

The solution of (Ω) that we have constructed is therefore

$$F(z, \lambda) = \exp z(\tfrac{1}{\lambda} - 1)V + \bar{z}(\lambda - 1)c(V),$$

and the corresponding harmonic map is

$$\phi(z) = F(z, -1) = \exp -2zV - 2\bar{z}c(V).$$

This is the "trivial harmonic map" of Chapter 10.

This calculation shows that an element V of $\mathbf{g} \otimes \mathbf{C}$ is of type 0 if and only if $[V, c(V)] = 0$.

(2) Harmonic maps of finite uniton number are usually *not* of finite type. However, from the Weierstrass formulae of Chapter 22, we can identify

some special examples which have both finite uniton number and finite type. Let us consider first the simplest such example, namely the identity map $S^2 \to S^2$. From section I of Chapter 22, this corresponds to the following solution of (Ω):

$$F(z, \lambda) = \left[\exp\left(\begin{matrix} 0 & 0 \\ \lambda^{-1}z & 0 \end{matrix}\right)\right]_u.$$

We claim that the composition $S^2 \to S^2 \subseteq SU_2$ is a harmonic map of type 1, with initial condition

$$V(\lambda) = \tfrac{1}{\lambda}\left(\begin{matrix} 0 & 0 \\ 1 & 0 \end{matrix}\right).$$

Although this appears to be a straightforward example of the previous theory, we cannot conclude that we have a map of type 1 until we verify that $X = F^{-1}VF$ is of the form $X = X_{-1}(1 - \lambda^{-1}) + X_0 + X_1(1 - \lambda)$.

From Chapter 22, we have

$$F = \left(\begin{matrix} 1 & \\ & \lambda \end{matrix}\right)^{-1}(\pi + \lambda\pi^{\perp}) = \left(\begin{matrix} 1 & \\ & \lambda \end{matrix}\right)^{-1}\psi\left(\begin{matrix} 1 & \\ & \lambda \end{matrix}\right)\psi^{-1}$$

where $\pi(z) : \mathbf{C}^2 \to \mathbf{C}^2$ is projection on the line $\mathbf{C}\left(\begin{matrix} 1 \\ z \end{matrix}\right)$, and

$$\psi(z) = \frac{1}{1 + |z|^2}\left(\begin{matrix} 1 & -\bar{z} \\ z & 1 \end{matrix}\right)$$

(cf. the discussion of harmonic sequences in Chapter 21). This gives

$$X(z, \lambda) = \frac{1}{(1 + |z|^2)^2}\left\{\lambda^{-1}\left(\begin{matrix} -\bar{z} & -\bar{z}^2 \\ 1 & \bar{z} \end{matrix}\right) + \left(\begin{matrix} \bar{z}(1 - |z|^2) & 2\bar{z}^2 \\ 2|z|^2 & \bar{z}(|z|^2 - 1) \end{matrix}\right) + \lambda\left(\begin{matrix} z|z|^2 & -\bar{z}^2 \\ |z|^4 & -z|z|^2 \end{matrix}\right)\right\}.$$

Thus, we have indeed got a harmonic map of type 1.

More generally, if N is any nilpotent matrix, then the formula $F(z, \lambda) = [\exp \lambda^{-1}zN]_u$ defines a solution of (Ω) which corresponds to a harmonic map of finite type.

III. An alternative version.

There is an alternative version of the theory of section I, based on (Λ) instead of (Ω). For this, we use the alternative version

$$\Lambda G^c = \Lambda G \Lambda_+^{\hat{N}} G^c$$

of the Iwasawa decomposition. Here, $G^c = GAN = G\hat{N}$ is the Iwasawa decomposition of G^c (as in Chapter 3), and $\Lambda_+^{\hat{N}}G^c = \{\gamma \in \Lambda_+G^c \mid \gamma(0) \in \hat{N}\}$. We use the notation $\gamma = \gamma_1\gamma_2$ for the factorization of a loop γ.

To find the projection π_1 onto the first summand of

$$\Lambda\mathbf{g} \otimes \mathbf{C} = (\Lambda\mathbf{g}) \oplus (\Lambda_+^{\hat{n}}\mathbf{g} \otimes \mathbf{C})$$

we write $f(\lambda) = \sum_{i\in\mathbf{Z}} A_i\lambda^i$, $f_+(\lambda) = \sum_{i>0} A_i\lambda^i$, and $f_-(\lambda) = \sum_{i<0} A_i\lambda^i$. We write $A_0 = A_0' + A_0''$ for the decomposition of A_0 with respect to $\mathbf{g} \otimes \mathbf{C} = \mathbf{g} \oplus \hat{n}$. The identity

$$f = (f_- + c(f_-) + A_0') + (f_+ - c(f_-) + A_0'')$$

shows that π_1 is given by $\pi_1(f) = f_- + c(f_-) + A_0'$.

Taking J_1, J_2 as before, and $X = \sum_{i=-d}^{d} X_i\lambda^i$, we obtain

$$\pi_1(J_1)_X = \pi_1(\lambda^{d-1}\sum_{i=-d}^{d} X_i\lambda^i)$$

$$= \tfrac{1}{\lambda}X_{-d} + c(\tfrac{1}{\lambda}X_{-d}) + X'_{-(d-1)}$$

$$\pi_1(J_2)_X = \pi_1(\sqrt{-1}\lambda^{d-1}\sum_{i=-d}^{d} X_i\lambda^i)$$

$$= \tfrac{1}{\lambda}\sqrt{-1}X_{-d} + c(\tfrac{1}{\lambda}\sqrt{-1}X_{-d}) + (\sqrt{-1}X_{-(d-1)})'.$$

Hence

$$F^{-1}F_z = \tfrac{1}{\lambda}X_{-d} + \tfrac{1}{2}(X'_{-(d-1)} - \sqrt{-1}(\sqrt{-1}X_{-(d-1)})') = \text{ linear in } \tfrac{1}{\lambda}$$

$$F^{-1}F_{\bar{z}} = \lambda c(X_{-d}) + \tfrac{1}{2}(X'_{-(d-1)} + \sqrt{-1}(\sqrt{-1}X_{-(d-1)})') = \text{ linear in } \lambda$$

and so F is a solution of (Λ). We have

$$F(z,\lambda) = (\exp s(J_1)_V + t(J_2)_V)_1$$

$$= (\exp s(\lambda^{d-1}V) + t(\sqrt{-1}\lambda^{d-1}V))_1$$

$$= (\exp z\lambda^{d-1}V(\lambda))_1.$$

The differential equations for X here are:

$$X_z = [X, \tfrac{1}{\lambda}X_{-d} + \tfrac{1}{2}(X'_{-(d-1)} - \sqrt{-1}(\sqrt{-1}X_{-(d-1)})')]$$

$$X_{\bar{z}} = [X, \lambda c(X_{-d}) + \tfrac{1}{2}(X'_{-(d-1)} + \sqrt{-1}(\sqrt{-1}X_{-(d-1)})')].$$

The harmonic maps constructed this way are again the harmonic maps of finite type. To see this, we observe that a harmonic map ϕ arises from the above construction if and only if there exists some $V(\lambda) = \sum_{i=-d}^{d} V_i \lambda^i$ such that the map $F(z, \lambda) = (\exp z\lambda^{d-1} V(\lambda))_1$ satisfies the following properties:

(i) F is a solution of (Λ) corresponding to ϕ, i.e., $\phi(z) = F(z, -1)F(z, 1)^{-1}$, and

(ii) $F^{-1}VF$ is of the form $\sum_{i=-d}^{d} X_i \lambda^i$.

Using the fact that $(\exp z\lambda^{d-1} V(\lambda))_1 = (\exp z\lambda^{d-1} V(\lambda))_u g(z)$ for some $g : \mathbf{C} \to G$, it may be verified that this characterization of ϕ is equivalent to condition (HF2) of section I.

Let us investigate the above differential equations more explicitly when $G = U_n$, $G^c = GL_n\mathbf{C}$. The involution $c : GL_n\mathbf{C} \to GL_n\mathbf{C}$ is given by $c(X) = -X^*$. The (finite dimensional) Iwasawa decomposition is

$$\mathbf{gl}_n\mathbf{C} = \mathbf{u}_n \oplus \hat{\mathbf{n}}$$

where $\hat{\mathbf{n}}$ is the space of upper triangular complex matrices whose diagonal entries are real. Let us compute the decomposition $X = X' + X''$ in this case. We write

$$X = u(X) + d(X) + l(X)$$

where $u(X), d(X), l(X)$ are (respectively) the upper triangular, diagonal, lower triangular parts of X. From the identity

$$X = (l(X) - l(X)^* + \sqrt{-1}\,\text{Im}\,d(X)) + (u(X) + l(X)^* + \text{Re}\,d(X))$$

we see that $X' = l(X) - l(X)^* + \sqrt{-1}\,\text{Im}\,d(X)$. From this we have

$$X' - \sqrt{-1}(\sqrt{-1}\,X)'$$
$$= l(X) - l(X)^* + \sqrt{-1}\,\text{Im}\,d(X) - \sqrt{-1}(\sqrt{-1}\,l(X) + \sqrt{-1}\,l(X)^* + \sqrt{-1}\,\text{Re}\,d(X))$$
$$= 2l(X) + d(X),$$

and so the ordinary differential equations in this example are

$$X_z = [X, \tfrac{1}{\lambda}X_{-d} + l(X_{-(d-1)}) + \tfrac{1}{2}d(X_{-(d-1)})]$$
$$X_{\bar{z}} = [X, -\lambda(X_{-d})^* - l(X_{-(d-1)})^* - \tfrac{1}{2}d(X_{-(d-1)})^*].$$

Example:

Example (2) of section II, with initial condition

$$V(\lambda) = \tfrac{1}{\lambda}\begin{pmatrix} 0 & 0 \\ 1 & 0 \end{pmatrix},$$

gives rise to the following solution of (Λ):

$$F(z, \lambda) = \left[\exp \begin{pmatrix} 0 & 0 \\ \lambda^{-1} z & 0 \end{pmatrix} \right]_1 = \begin{pmatrix} 1 & \\ & \lambda \end{pmatrix}^{-1} \psi \begin{pmatrix} 1 & \\ & \lambda \end{pmatrix}.$$

The corresponding map $X = F^{-1} V F$ is

$$X(z, \lambda) = \frac{1}{1 + |z|^2} \left\{ \lambda^{-1} \begin{pmatrix} 0 & 0 \\ 1 & 0 \end{pmatrix} + \begin{pmatrix} \bar{z} & 0 \\ 0 & -\bar{z} \end{pmatrix} + \lambda \begin{pmatrix} 0 & -\bar{z}^2 \\ 0 & 0 \end{pmatrix} \right\}.$$

Thus, the formulae are somewhat simpler in this new version. We note further (from the formula for F) that

$$F^{-1} F_z = \frac{1}{1 + |z|^2} \left\{ \lambda^{-1} \begin{pmatrix} 0 & 0 \\ 1 & 0 \end{pmatrix} + \frac{1}{2} \begin{pmatrix} \bar{z} & 0 \\ 0 & -\bar{z} \end{pmatrix} \right\}.$$

The right hand side is equal to $\lambda^{-1} X_{-1} + l(X_0) + \frac{1}{2} d(X_0)$, as predicted by the previous paragraph.

IV. Primitive maps of finite type.

Let $\sigma : G \to G$ be an automorphism of order k, with $k > 2$. Let $K = G_\sigma$, i.e., the fixed point set of σ. The "twisted" version of the theory of section III leads to the notion of primitive maps of finite type. For this we need the twisted Iwasawa decomposition

$$(\Lambda G^c)_\sigma = (\Lambda G)_\sigma (\Lambda_+^M G^c)_\sigma$$

where $K^c = KM$ is the Iwasawa decomposition of the group K^c. (For the twisted Iwasawa decomposition, see Theorem 2.3 of Dorfmeister $et\ al.$ [in press]; we use a special case in Chapter 13.) In this section we use the notation $\gamma = \gamma_1 \gamma_2$ for the factorization of a loop $\gamma \in (\Lambda G^c)_\sigma$.

Using the method of section III, we obtain a solution of (Λ_σ) for any initial condition $V(\lambda) = \sum_{i=-d}^{d} V_i \lambda^i$ in $(\Lambda \mathbf{g})_\sigma \otimes \mathbf{C}$ of finite type. Following Burstall and Pedit [1994], we say that the corresponding map $\phi : \mathbf{C} \to G/K$ is a $primitive\ map\ of\ finite\ type$ (and of $type\ d$). Note that we must have $V_i \in \mathbf{g}_i$, where $\mathbf{g}_i$ is the (ω^i)-eigenspace of σ on $\mathbf{g} \otimes \mathbf{C}$, as in Chapter 21.

Such maps ϕ may be characterized in terms of F or X. Before giving these characterizations, we recall some further notation from Chapter 21. Let ϕ be a primitive map. If $\phi = [\psi]$ for some $\psi : \mathbf{C} \to G$, then $\psi^{-1} \psi_z = A + B$, where A, B are (respectively) the components of $\psi^{-1} \psi_z$ in $\mathbf{g}_0, \mathbf{g}_{-1}$. We then have:

(PF1)   Let $\phi = [\psi] : \mathbf{C} \to G/K$ be a primitive map. Let d be a non-negative integer. Then ϕ has type d if and only if there exists some $V(\lambda) = \sum_{i=-d}^{d} V_i \lambda^i \in (\Lambda \mathbf{g})_\sigma \otimes \mathbf{C}$ such that

$$X_{-d} = B, \quad \tfrac{1}{2}(X'_{-(d-1)} - i(iX_{-(d-1)})') = A,$$

where $X = \sum_{i=-d}^{d} X_i \lambda^i : \mathbf{C} \to (\Lambda\mathbf{g})_\sigma \otimes \mathbf{C}$ is the solution of

$$X_z = [X, \tfrac{1}{\lambda}X_{-d} + \tfrac{1}{2}(X'_{-(d-1)} - i(iX_{-(d-1)})')]$$
$$X_{\bar{z}} = [X, \lambda c(X_{-d}) + \tfrac{1}{2}(X'_{-(d-1)} + i(iX_{-(d-1)})')]$$

with $X(0) = V$. (Since $X_d = B$ takes values in $\mathbf{g}_{-1}$, we necessarily have $d \equiv 1 \bmod k$.)

(PF2)   Let $\phi : \mathbf{C} \to G/K$ be a primitive map. Let d be a non-negative integer. Then ϕ has type d if and only if there exists some $V(\lambda) = \sum_{i=-d}^{d} V_i \lambda^i \in (\Lambda\mathbf{g})_\sigma \otimes \mathbf{C}$ such that $F(z, \lambda) = (\exp z\lambda^{d-1}V(\lambda))_1$ is a solution of (Λ_σ) corresponding to ϕ, and $X = F^{-1}VF$ is of the form $X = \sum_{i=-d}^{d} X_i \lambda^i$.

From a primitive map of finite type, we obtain harmonic maps into symmetric spaces by the method of Chapter 21, section IV. These harmonic maps are not necessarily of finite type. Hence, primitive maps of finite type provide a significant generalization of harmonic maps of finite type.

We have insisted that σ be an automorphism of order *greater than two*, so far. If σ has order two, and we consider harmonic maps $\mathbf{C} \to G/K$ instead of primitive maps, then there are two possible definitions of "finite type". On the one hand, we may immerse G/K into G via the Cartan embedding, and then use (HF1) or (HF2). On the other hand, we may modify (PF1) or (PF2), replacing "primitive" by "harmonic". It can be shown (by using the method of 5.4 of Burstall and Pedit [1994]) that these definitions are equivalent. This remains true[7] if we impose the reality condition.

Bibliographical comments.

See the comments at the end of Chapter 25.

---

[7]Note, however, that the definition of finite type in Burstall and Pedit [1994] requires X to be based, as well as real. With this more restrictive definition it is clear that (HF2) implies (PF2), but not vice versa.

Chapter 25: Application: Harmonic maps from T^2 to S^2

Harmonic maps $\mathbf{C} \to G$ or G/K of finite type, like harmonic maps of finite uniton number, are rather special. However, all harmonic maps which extend to $S^2 = \mathbf{C} \cup \infty$ are of finite uniton number, and these constitute a very significant class of examples. Other significant examples are the harmonic maps which are doubly periodic, i.e., which factor through a torus $T^2 = \mathbf{C}/L$, where L is a lattice of rank 2 in $\mathbf{C}$. For certain G/K, it turns out that *all harmonic maps $T^2 \to G/K$ are either of finite uniton number or of finite type* (or, more generally, obtained from primitive maps of finite type by the method of Chapter 21). The fundamental example of this phenomenon is provided by $G/K = SU_2/S^1 = S^2$. In this chapter we sketch the theory for this case.

I. Conformal and nowhere conformal maps $T^2 \to S^2$.

Let $\sigma : SU_2 \to SU_2$ be the involution given by conjugation by $\begin{pmatrix} 1 & \\ & -1 \end{pmatrix}$.
Then $(SU_2)_\sigma = S^1$ and the symmetric space $SU_2/(SU_2)_\sigma$ may be identified with S^2.

Let $\phi = [\psi] : \mathbf{C} \to SU_2/S^1$ be a harmonic map, with $\psi : \mathbf{C} \to SU_2$. From Chapter 18 we know that $\psi(z) = F(z, 1)$, for some solution $F : \mathbf{C} \to (\Lambda SU_2)_\sigma$ of (Λ_σ).

We shall assume that ϕ factors through $T^2 = \mathbf{C}/L$. The main difficulty we face is that F does not necessarily factor through T^2. (This is in contrast with the situation for maps which extend to $S^2 = \mathbf{C} \cup \infty$ – see the bibliographical comments for Chapters 16-17.) Indeed, it can be shown that (there exists a solution F such that) F factors through T^2 if and only if ϕ has finite uniton number. In that case the theory of Chapter 20 applies, and ϕ must be holomorphic or anti-holomorphic (in other words ϕ is a meromorphic function on T^2, i.e., an elliptic function).

We shall begin with some general remarks on (smooth) maps $\phi = [\psi] : \mathbf{C} \to S^2$. Such a map is said to be *conformal* if $\langle d\phi(\frac{\partial}{\partial z}), d\phi(\frac{\partial}{\partial z}) \rangle = 0$, where $\langle \, , \, \rangle$ is the $\mathbf{C}$-linear extension of the standard Riemannian metric on S^2. This condition is equivalent to the condition $\langle \pi(\psi^{-1}\psi_z), \pi(\psi^{-1}\psi_z) \rangle = 0$, where $\langle \, , \, \rangle$ now denotes the Killing form of $\mathfrak{sl}_2\mathbf{C}$, and where $\pi(\psi^{-1}\psi_z)$ means the component of $\psi^{-1}\psi_z$ in the (-1)-eigenspace of σ on $\mathfrak{sl}_2\mathbf{C}$.

The relevance of the torus T^2 comes from the following well known fact:

Proposition. *If $\phi : \mathbf{C} \to S^2$ is a harmonic map which factors through the torus T^2, then either $\langle d\phi(\frac{\partial}{\partial z}), d\phi(\frac{\partial}{\partial z}) \rangle$ is identically zero or it is never zero. In other words, either ϕ is conformal or it is nowhere conformal.*

Sketch of the proof. It follows from the harmonic map equation that

$$(Z, W) \mapsto \langle d\phi(Z), d\phi(W) \rangle$$

defines a holomorphic section of the bundle $(TT^2)^* \otimes (TT^2)^*$ (see (4.7) of Eells and Lemaire [1988]). This bundle is (holomorphically) trivial, as T^2 is a complex Lie group. Since T^2 is also compact, any holomorphic section is either identically zero or never zero. ∎

These two kinds of behaviour can be expressed in more familiar form:

Proposition.

(1) A harmonic map $\mathbf{C} \to S^2$ is conformal if and only if it is holomorphic or anti-holomorphic.

(2) A harmonic map $\mathbf{C} \to S^2$ is nowhere conformal if and only if it corresponds to a solution of the 2DTL for SU_2 (via the construction of Chapter 21).

Proof. We have $\phi = [\psi]$ where

$$\pi(\psi^{-1}\psi_z) = \begin{pmatrix} 0 & u \\ v & 0 \end{pmatrix}.$$

Since the Killing form of $\mathbf{sl}_2\mathbf{C}$ is given (up to a constant) by $\langle A, B \rangle = \text{trace}\, AB$, we have

$$\langle \pi(\psi^{-1}\psi_z), \pi(\psi^{-1}\psi_z) \rangle = \text{constant} \times uv.$$

The map ϕ is conformal if and only if u or v is identically zero; this means that ϕ is holomorphic or anti-holomorphic.

The map ϕ is nowhere conformal if and only if both u and v are never zero. By the proof of the previous proposition, we can assume that uv is constant, and in fact that $uv = 1$. On replacing ψ by

$$\tilde{\psi} = \psi \begin{pmatrix} e^{\sqrt{-1}\,p} & \\ & e^{-\sqrt{-1}\,p} \end{pmatrix},$$

where p is a positive real-valued function on $\mathbf{C}$ (to be specified later), we obtain

$$\pi(\tilde{\psi}^{-1}\tilde{\psi}_z) = \begin{pmatrix} 0 & ue^{-2\sqrt{-1}\,p} \\ u^{-1}e^{2\sqrt{-1}\,p} & 0 \end{pmatrix}.$$

(This operation does not change ϕ.) If $u = be^{\sqrt{-1}\,a}$, with a, b real and $b > 0$, then we choose $p = a/2$. We obtain

$$\pi(\tilde{\psi}^{-1}\tilde{\psi}_z) = \begin{pmatrix} 0 & b \\ b^{-1} & 0 \end{pmatrix},$$

with $b > 0$. Thus, condition (T) of Chapter 13 is satisfied, so $\tilde{\psi}$ gives a solution of the 2DTL. ∎

II. Nowhere conformal maps $T^2 \to S^2$ are of finite type.

To prove that a harmonic (or primitive) map ϕ is of finite type, it is necessary to produce a "vector field" $X : \mathbf{C} \to (\Lambda \mathbf{g})_\sigma \otimes \mathbf{C}$ satisfying certain conditions. This vector field can be interpreted as a special kind of tangent vector at $F : \mathbf{C} \to (\Lambda G)_\sigma$, where F is a solution of (Λ_σ) corresponding to ϕ. To see this, consider any (smooth) family F_t of solutions to (Λ_σ), with $F_0 = F$. It is reasonable to interpret the map

$$X = F^{-1} \tfrac{d}{dt} F_t|_0 : \mathbf{C} \to (\Lambda \mathbf{g})_\sigma$$

as a tangent vector to F. Observe that

$$X_z = -F^{-1} F_z F^{-1} \tfrac{d}{dt} F_t|_0 + F^{-1} \tfrac{d}{dt} (F_t)_z|_0$$

and

$$\tfrac{d}{dt} F_t^{-1} (F_t)_z|_0 = -F^{-1} \tfrac{d}{dt} F_t|_0 F^{-1} F_z + F^{-1} \tfrac{d}{dt} (F_t)_z|_0.$$

Subtracting these equations, we obtain

$$X_z = [X, F^{-1} F_z] + \tfrac{d}{dt} F_t^{-1} (F_t)_z|_0.$$

Motivated by this, and following Pinkall and Sterling [1989]; Ferus *et al.* [1992]; Burstall *et al.* [1993], we say that a map $X : \mathbf{C} \to (\Lambda \mathbf{g})_\sigma \otimes \mathbf{C}$ is a *Jacobi field* (for F) if it satisfies the system

(J) $X_z - [X, F^{-1} F_z] = \text{linear in } \tfrac{1}{\lambda}$

$X_{\bar z} - [X, F^{-1} F_{\bar z}] = \text{linear in } \lambda.$

We say that X is a *Killing field* (for F) if it satisfies the system

(K) $X_z - [X, F^{-1} F_z] = 0$

$X_{\bar z} - [X, F^{-1} F_{\bar z}] = 0.$

Thus, to prove that a map is of finite type, we need a Killing field which is a finite polynomial in λ and λ^{-1}.

The strategy for finding such a polynomial Killing field depends on the following result (see §4 of Ferus *et al.* [1992]):

Lemma. *Let $Y = \sum_{i \ge -N} Y_i \lambda^i : T^2 \to (\Lambda \mathbf{g})_\sigma \otimes \mathbf{C}$ be a formal Killing field for F, with $N \ge 0$. Then there exists a polynomial Killing field X for F.*

Proof. We have

$$F^{-1} F_z = A + \tfrac{1}{\lambda} B$$
$$F^{-1} F_{\bar z} = C + \lambda D$$

as usual. The hypothesis on Y means that

$$Y_z - [Y, A + \tfrac{1}{\lambda}B] = 0$$
$$Y_{\bar{z}} - [Y, C + \lambda D] = 0$$

in the sense that the coefficients of λ^i in each equation are zero, for all i.

Let $Y_{\leq 0} = \sum_{i=-N}^{0} Y_i \lambda^i$ and let $Y_{>0} = \sum_{i>0} Y_i \lambda^i$. Consider the linear operators

$$L' : Y \longmapsto W' = (Y_{\leq 0})_z - [Y_{\leq 0}, A + \tfrac{1}{\lambda}B]$$
$$L'' : Y \longmapsto W'' = (Y_{\leq 0})_{\bar{z}} - [Y_{\leq 0}, C + \lambda D].$$

(If $Y \in \text{Ker}\, L' \cap \text{Ker}\, L''$, then $Y_{\leq 0}$ is a polynomial Killing field.) We claim that W', W'' satisfy the following equations:

(i) $W' = (Y_0)_z - [Y_0, A]$, $W'' = -\lambda[Y_0, D]$.

(Proof: Substituting $Y = Y_{\leq 0} + Y_{>0}$ into the definition of Killing field, we obtain $W' = (Y_{\leq 0})_z - [Y_{\leq 0}, A + \tfrac{1}{\lambda}B] = -(Y_{>0})_z + [Y_{>0}, A + \tfrac{1}{\lambda}B]$. Comparing coefficients of λ^i, we see that W' has only a constant term, which must be $(Y_0)_z - [Y_0, A]$. A similar argument with W'' gives the second formula.)

(ii) $W'_{\bar{z}} - W''_z = [W', C + \lambda D] - [W'', A + \tfrac{1}{\lambda}B]$.

(Proof: This follows directly from the definitions of W', W'', together with the zero-curvature equation for $A + \tfrac{1}{\lambda}B$, $C + \lambda D$.)

Substituting (i) into (ii), and comparing coefficients of λ^0, we find that Y_0 satisfies a second order elliptic partial differential equation. Since T^2 is compact, we deduce that the images of L' and L'' must be finite dimensional.

On the other hand, we obtain an *infinite* sequence of linearly independent formal Killing fields by taking $\lambda^{-k_i}Y$ with $k_1 < k_2 < \ldots$. It follows that some non-trivial linear combination of these must be in the intersection of the kernels of L' and L''. This gives the required polynomial Killing field X. ∎

We shall apply this to the case of a nowhere conformal harmonic map $\phi : T^2 \to S^2$. From section I, we know that ϕ corresponds to a solution w_0, w_1 of the 2DTL, or a periodic harmonic sequence f_0, f_1 (of period 2). This allows us to write $\phi = [\psi]$ where

$$\psi = PD^{-1}, \quad P = (\,\hat{f}_0 \quad \hat{f}_1\,), \quad D = \begin{pmatrix} \|\hat{f}_0\| & \\ & \|\hat{f}_1\| \end{pmatrix}$$

as in Chapter 21. We have

$$\psi^{-1}\psi_z = \begin{pmatrix} (w_0)_z & \\ & (w_1)_z \end{pmatrix} + \begin{pmatrix} & W_{0,1} \\ W_{1,0} & \end{pmatrix}$$

where $w_i = \log \|\hat{f}_i\|$, and this can be re-written in the form

$$\psi^{-1}\psi_z = D^{-1}D_z + D \begin{pmatrix} & 1 \\ 1 & \end{pmatrix} D^{-1}.$$

The main task now is to find a formal Killing field $Y = \sum_{i \geq 1} Y_i \lambda^i$ such that

$$Y_1 = D \begin{pmatrix} & 1 \\ 1 & \end{pmatrix} D^{-1}, \quad Y_2 = 2D^{-1}D_z.$$

This may be done by using a technique from the theory of asymptotic expansions; we refer to §4 of Ferus *et al.* [1992] for the details. The lemma then gives a polynomial Killing field $X = \sum_{i=-d}^{d} X_i \lambda^i$ for some d. From the proof of the lemma, we have

$$X_{-d} = Y_1, \quad X_{-(d-1)} = Y_2$$

(if we choose the differences $k_i - k_{i-1}$ sufficiently large). This means that ψ is of finite type in the sense of section IV of Chapter 24. We conclude:

Theorem (Pinkall and Sterling [1989]). *Any nowhere conformal harmonic map $T^2 \to S^2$ is of finite type.* ∎

III. Generalizations.

It follows from the formulae in section II that a nowhere conformal map $\phi : \mathbf{C} \to S^2 \subseteq SU_2$ has the property that $\phi^{-1}\phi_z$ takes values in a single orbit of the adjoint action of $SL_2\mathbf{C}$ on $\mathrm{sl}_2\mathbf{C}$. This is the key to the existence of a formal Killing field in the previous section, and leads to various generalizations.

The first result in this direction is:

Theorem (Burstall *et al.* [1993]). *Let G be a compact semisimple Lie group (or $G = U_n$). Let $\phi : \mathbf{C} \to G$ be a harmonic map such that*

(i) ϕ factors through a torus $T^2 = \mathbf{C}/L$

(ii) $\phi^{-1}\phi_z$ takes values in an orbit $\mathrm{Ad}(G^c)E$, for some semisimple element $E \in \mathbf{g} \otimes \mathbf{C}$.

Then ϕ is of finite type (i.e., ϕ satisfies conditions (HF1) and (HF2) of Chapter 24). ∎

(An element E of $\mathbf{g} \otimes \mathbf{C}$ is said to be semisimple if the linear transformation $\mathrm{ad}\, E$ is diagonalizable.)

If ϕ is a nowhere conformal harmonic map from a torus into a symmetric space of rank 1, condition (ii) is satisfied, and so any such map must be of finite type. (This includes the situation of maps into S^2.)

A similar generalization is possible in the case of primitive maps $\phi : \mathbf{C} \to G/K$. The 2DTL provides an example, i.e., the case of a primitive map associated to a periodic harmonic sequence:

Theorem (Bolton *et al.* **[1995]).** *Let* $\phi : \mathbf{C} \to SU_{n+1}/T_{n+1}$ *be the primitive map associated to a periodic harmonic sequence, as in Chapter 21. Assume that* ϕ *factors through a torus* $T^2 = \mathbf{C}/L$. *Then* ϕ *is of finite type (i.e.,* ϕ *satisfies conditions (PF1) and (PF2) of Chapter 24).* ∎

These results indicate that harmonic and primitive maps of finite type constitute a non-trivial class of examples. Although the role of such maps is still rather mysterious, the situation for maps into S^n or $\mathbf{C}P^n$ has been clarified considerably in Burstall [1995]; McIntosh [1995]. In the case of $\mathbf{C}P^n$, this depends on a generalization of the dichotomy of conformal and nowhere conformal harmonic maps $T^2 \to S^2 = \mathbf{C}P^1$. For $n \geq 2$, we may consider the harmonic sequence $f_0, f_1, \ldots$ of a harmonic map $f = f_0 : T^2 \to \mathbf{C}P^n$; if $f_0, f_1, \ldots, f_{i-1}$ are mutually orthogonal, but $f_0, f_1, \ldots, f_i$ are not, then we say that f is *i-orthogonal* (Bolton *et al.* [1995]).

At one extreme are the $(n+1)$-orthogonal harmonic maps of Chapter 21 – where the harmonic sequence is either finite or periodic. If the harmonic sequence is finite, f is called *superminimal*, and is a map of finite uniton number. (Moreover, the harmonic sequence corresponds to a primitive map of finite uniton number.) If the harmonic sequence is periodic, f is called *superconformal*, and we have just seen that the harmonic sequence corresponds to a primitive map of finite type.

At the other extreme, we have 2-orthogonal harmonic maps; these are precisely the nowhere conformal harmonic maps (here we use the fact that the domain is a torus). These maps are also of finite type, by the first theorem above.

It is shown in Burstall [1995] that the intermediate cases, i.e., k-orthogonal harmonic maps with $3 \leq k \leq n$, also arise from primitive maps of finite type. Thus, *every* harmonic map from T^2 to $\mathbf{C}P^n$ is either of finite uniton number, or of finite type, or arises from a primitive map of finite type. A similar result holds for maps into S^n.

(In the case $n = 1$, the above terminology is rather unfortunate: All harmonic maps are 2-orthogonal, superminimal is the same as conformal, and superconformal is the same as nowhere conformal!)

Bibliographical comments for Chapters 24-25.

Harmonic maps of finite type were introduced and studied in Pinkall and Sterling [1989]; Ferus *et al.* [1992]; Burstall *et al.* [1993]. Primitive maps of finite type were introduced in Burstall and Pedit [1994]. Explicit numerical computations are possible for such maps, by solving the appropriate pair of ordinary differential equations. Some examples are depicted in Pinkall and Sterling [1989].

The fact that a nowhere conformal harmonic map $T^2 \to S^2$ is of finite type was proved in Pinkall and Sterling [1989]. This result was generalized to the case of maps $T^2 \to S^4$ in Ferus *et al.* [1992], by making use of a relationship with the 2DTL for the group SO_5. The general case of doubly periodic nowhere conformal harmonic maps into symmetric spaces appeared in Burstall *et al.* [1993]. Doubly periodic solutions of the general 2DTL were treated in Bolton *et al.* [1995].

For results on harmonic maps of tori into $\mathbb{C}P^n$, see Jensen and Liao [1995]; Burstall [1995]; McIntosh [1995].

Chapter 26: Epilogue

I. A brief review.

From the point of view of harmonic maps, this book has addressed three principal questions:

(1) The harmonic map equation is concerned with the differential geometry of finite dimensional manifolds. What does it have to do with infinite dimensional Lie groups?

(2) Given that there is a connection between harmonic maps and infinite dimensional Lie groups, do these infinite dimensional Lie groups tell us anything new? Is this connection useful?

(3) What does this have to do with the theory of integrable systems?

As motivation, we considered a famous example, the Toda lattice. The one-dimensional finite open Toda lattice (the 1DTL) is the following system of second order ordinary differential equations[8] for $q_1, \ldots, q_n : \mathbf{R} \to \mathbf{R}$:

$$\ddot{q}_1 = e^{q_2 - q_1}$$
$$\ddot{q}_i = e^{q_{i+1} - q_i} - e^{q_i - q_{i-1}} \quad i = 2, \ldots, n-1$$
$$\ddot{q}_n = \qquad\qquad e^{q_n - q_{n-1}}.$$

In Chapters 4-8 we discussed this system in some detail. Lie groups (finite dimensional, in this case) enter the picture because the system may be written in matrix form as a Lax equation

$$\dot{L} = [L, M].$$

For a given initial condition $L(0) = V$, the unique solution L is given by the explicit formula $L = U^{-1}VU$, where

$$U(t) = [\exp tV]_1.$$

Here, $\exp tV = [\exp tV]_1 [\exp tV]_2$ is a certain matrix factorization.

The two-dimensional periodic Toda lattice (the 2DTL) is the following system of second order partial differential equations for $w_i : \mathbf{C} \to \mathbf{R}$:

$$(w_i)_{z\bar{z}} = e^{w_{i+1} - w_i} - e^{w_i - w_{i-1}} \quad i \in \mathbf{Z}, w_i = w_{i+n}.$$

This was introduced in Chapters 9-10 and then studied in more detail in Chapters 13-15. The symmetry group is now infinite dimensional, and there is no easy way to find solutions.

---

[8]The constants have been modified in both the 1DTL and the 2DTL in this chapter. The versions used in the rest of the book are slightly different, in order to simplify the matrix forms of the equations.

Our treatment of the Toda lattice was not intended to be a thorough exposition; it was a convenient way of introducing various ideas which are useful for studying harmonic maps (our main goal). For example, we did not mention the one-dimensional *periodic* Toda lattice, which is extremely interesting as – despite apparent similarities with the finite open Toda lattice – its symmetry group is infinite dimensional (see Adler and van Moerbeke [1980a; 1980b]; Arnold and Novikov [1994]). We restricted our attention primarily to the Lie group $SL_n\mathbf{C}$, although there is a beautiful and intricate generalization valid for general semisimple Lie groups (Kostant [1979]).

The harmonic map equation, for maps $\phi : \mathbf{C} \to G$, where G is a compact matrix group, is the second order partial differential equation

$$(\phi^{-1}\phi_{\bar{z}})_z + (\phi^{-1}\phi_z)_{\bar{z}} = 0.$$

The relevance of loop groups to this equation was discussed in Chapters 9-10, and in Chapters 16-25 techniques from loop group theory were used to investigate harmonic maps in detail. The one-dimensional and two-dimensional Toda lattices both play significant roles in this investigation. Solutions of the 2DTL correspond to families of special harmonic maps (Chapter 21), and a generalization of the 1DTL gives rise to harmonic maps via the "1 + 1 = 2" principle (Chapters 23-24). In particular, both partial differential equations in this book are examples of a single general system. We call this system (Λ_σ).

II. The system (Λ_σ).

Let G be a compact Lie group, and let ΛG be the free loop group of G. Consider the following conditions on a map $F : \mathbf{C} \to \Lambda G$:

(Λ) $\qquad\qquad F^{-1}F_z = $ linear in $\frac{1}{\lambda}$

$\qquad\qquad\qquad F^{-1}F_{\bar{z}} = $ linear in λ

where λ is the "loop parameter". (*A priori*, $F^{-1}F_z$ is represented by a power series in λ, i.e., its Fourier series, so the requirement that this be linear is a non-trivial condition.) These two conditions are in fact equivalent, via the conjugation operator c of $\mathbf{g} \otimes \mathbf{C}$, but it is helpful to write down both of them.

Given a solution F of (Λ), let us write explicitly

$$F^{-1}F_z = A + \tfrac{1}{\lambda}B$$
$$F^{-1}F_{\bar{z}} = C + \lambda D,$$

where $A, B, C, D : \mathbf{C} \to \mathbf{g} \otimes \mathbf{C}$. We have $C = c(A)$, $D = c(B)$. It is easy to verify that the "zero-curvature equation"

$$(A + \tfrac{1}{\lambda}B)_{\bar{z}} - (C + \lambda D)_z = [A + \tfrac{1}{\lambda}B, C + \lambda D]$$

is satisfied. Conversely, if $A, B, C = c(A), D = c(B)$ are arbitrary (smooth) functions which satisfy this equation then there exists a solution F of (Λ) such that $F^{-1}F_z = A + \lambda^{-1}B$ and $F^{-1}F_{\bar{z}} = C + \lambda D$. The coefficients of $\lambda^{-1}, \lambda^0, \lambda^1$ give a collection of differential equations relating A, B, C, D. This (see Chapter 9) is the fundamental principle which relates loop group valued maps F with solutions A, B, C, D of a "finite dimensional" differential equation. The relevant differential equation in this case is equivalent to the harmonic map equation, but many other important differential equations arise in a similar way.

The system (Λ_σ) is obtained by a slight modification: We use the twisted loop group $(\Lambda G)_\sigma$ instead of ΛG, where $\sigma : G \to G$ is an automorphism of order k.

To describe this system in more detail, it is convenient to consider three separate cases:

(i) $k = 1$ (Chapters 9,16)

In this case, σ is the identity, and $(\Lambda G)_\sigma = \Lambda G$. The zero-curvature equation is equivalent to the equation for a harmonic map $\phi : \mathbf{C} \to G$, with

$$A = \tfrac{1}{2}\phi^{-1}\phi_z = -B$$
$$C = \tfrac{1}{2}\phi^{-1}\phi_{\bar{z}} = -D.$$

More precisely, one has:

(a) Let $\phi : \mathbf{C} \to G$, and let A, B, C, D be defined in terms of ϕ as above. If ϕ is harmonic, then $A + \lambda^{-1}B$ and $C + \lambda D$ satisfy the zero-curvature equation.

(b) Conversely, let $A, B, C, D : \mathbf{C} \to \mathbf{g} \otimes \mathbf{C}$ be functions such that $B = -A$, $C = c(A)$, $D = -C$. If $A + \lambda^{-1}B$ and $C + \lambda D$ satisfy the zero-curvature equation, then there exists a harmonic map $\phi : \mathbf{C} \to G$ which is related to A, B, C, D in the above manner.

(ii) $k = 2$ (Chapter 18)

In this case, σ is an involution on G. It determines a symmetric space G/K, where K is the fixed point set of σ. The zero-curvature equation is now equivalent to the equation for a harmonic map $\phi : \mathbf{C} \to G/K$. To obtain A, B, C, D in this case, one chooses a "lifting" or "framing" of ϕ, i.e., a map $\psi : \mathbf{C} \to G$ such that $\phi = [\psi]$. Then A, B, C, D are defined by

$$\psi^{-1}\psi_z = A + B$$
$$\psi^{-1}\psi_{\bar{z}} = C + D$$

where A, C take values in the $(+1)$-eigenspace of (the derivative of) σ on $\mathbf{g} \otimes \mathbf{C}$, and B, D take values in the (-1)-eigenspace. We then have:

(a) Let $\phi : \mathbf{C} \to G/K$, and let A, B, C, D be defined in terms of ψ as above. If ϕ is harmonic, then $A + \lambda^{-1}B$ and $C + \lambda D$ satisfy the zero-curvature equation.

(b) Conversely, let A, B, C, D be functions (of the above type) such that $C = c(A)$, $D = c(B)$. If $A + \lambda^{-1}B$ and $C + \lambda D$ satisfy the zero-curvature equation, then there exists a harmonic map $\phi : \mathbf{C} \to G$ which is related to A, B, C, D in the above manner.

The case $k = 1$ may in fact be expressed as a special case of this, by considering the Lie group G as the symmetric space $G \times G/\Delta$, where $\sigma(g_1, g_2) = (g_2, g_1)$ and Δ is the diagonal subgroup of $G \times G$. There is a canonical choice for ψ here, namely $\psi = (\phi, e)$, from which we recover the situation of (i).

(iii) $k > 2$ (Chapter 21)

A similar principle applies to the case $k > 2$. Let $\mathbf{g}_j$ denote the $(e^{2\pi\sqrt{-1}j/k})$-eigenspace of (the derivative of) σ on $\mathbf{g} \otimes \mathbf{C}$. For a map $\psi : \mathbf{C} \to G$, we define A, B, C, D by

$$\psi^{-1}\psi_z = A + B$$
$$\psi^{-1}\psi_{\bar{z}} = C + D$$

where A, C take values in $\mathbf{g}_0$, B takes values in $\mathbf{g}_{-1}$, and D takes values in $\mathbf{g}_1$.

The interpretation of the zero-curvature equation is more complicated in this case. Roughly, ψ corresponds to a *collection* of harmonic maps $\phi_\alpha : \mathbf{C} \to G/K_\alpha$, where $\{G/K_\alpha\}_\alpha$ is a collection of symmetric spaces associated to σ.

This is a generalization of the case $k = 2$, but there is a special feature in the present situation: For $k > 2$, the zero-curvature equation for $A + \lambda^{-1}B$ and $C + \lambda D$ is *equivalent* to the zero-curvature equation for $A + B$ and $C + D$. Geometrically, the latter equation corresponds to a simple condition – which we call (P) – on the derivative of the map $\phi = [\psi] : \mathbf{C} \to G/K$, where K is the fixed point set of σ. The homogeneous space G/K is a k-symmetric space, and a map ϕ which satisfies condition (P) is said to be *primitive*. This feature is not present in the case $k = 2$; in fact condition (P) is vacuous when $k = 2$.

The 2DTL fits into this scheme in the following way (Chapters 13, 21). Let $G = SU_{n+1}$, and let σ be the automorphism which is given by conjugation by a diagonal matrix whose diagonal entries are the $(n + 1)$-th roots of unity. Then solutions of the (periodic) 2DTL correspond to solutions of the system (Λ_σ) which satisfy an *additional* condition, which we call (T). Thus, solutions of the 2DTL correspond to certain primitive maps if $n > 1$, and to certain harmonic maps if $n = 1$. These maps have a simple characterization (Chapters 21,25): For $n > 1$, we obtain collections of $n + 1$ harmonic maps into $\mathbb{C}P^n$ which constitute a "periodic harmonic sequence"; for $n = 1$ we obtain "nowhere conformal" harmonic maps.

III. Applications of loop groups to harmonic maps.

The correspondence between harmonic maps $\phi = [\psi]$ and solutions F of (Λ_σ) introduces loop groups into the theory of harmonic maps. In the terminology of Uhlenbeck [1989] (for $k = 1$), F is called an extended harmonic map, or an extended solution. In the terminology of Burstall and Pedit [1994] (for $k \geq 2$), F is called an extended framing. Whereas ϕ is a function of $z \in \mathbb{C}$, the extended map F can be regarded as a function of $z \in \mathbb{C}$ and $\lambda \in S^1$. Given a solution F of (Λ_σ), a corresponding harmonic map $\phi = [\psi]$ is obtained by setting $\lambda = 1$, i.e., $\psi(z) = F(z, 1)$. (For $k = 1$, the formula is $\phi(z) = F(z, -1)$, because of the identification $G \cong G \times G / \Delta$.) On the other hand, the reverse construction (of F from ϕ) is non-trivial, and there is no simple formula. Thus, the "hidden" parameter λ may be expected to reveal intimate properties of the harmonic map ϕ – and it does.

Harmonic maps of finite uniton number: The simplest situation occurs when the Fourier series of $F(z, \lambda)$ has finitely many terms, i.e., it is "algebraic" in λ. The highest power of λ is called the uniton number; the corresponding harmonic maps are said to have finite uniton number. All harmonic maps ϕ which extend to the sphere $S^2 = \mathbb{C} \cup \infty$ are of this kind. The basic theory of such maps was established in Uhlenbeck [1989]. The loop group point of view was established in Segal [1989].

For $G = U_n$, the main results are:

(i) the maximum value of the uniton number is $n - 1$, and

(ii) a solution F of uniton number k is given by an explicit "Weierstrass formula" of the form

$$F(z, \lambda) = \left[\exp \begin{pmatrix} \lambda^{k_1} & \\ & \ddots & \\ & & \lambda^{k_n} \end{pmatrix} \right)^{-1} \sum_{i=0}^{k-1} C_i(z) \lambda^i \begin{pmatrix} \lambda^{k_1} & \\ & \ddots & \\ & & \lambda^{k_n} \end{pmatrix} \right]_1$$

where $C_0, \ldots, C_{k-1}$ are meromorphic matrix-valued functions, and $k = k_1 \geq \cdots \geq k_n = 1$. Similar results hold for harmonic maps $S^2 \to \mathrm{Gr}_k(\mathbb{C}^n)$. An introduction to these results was given in Chapters 16-20 and 22. For further details, and for results on more general compact Lie groups or symmetric spaces, see Burstall and Guest [preprint].

Our discussion was based on geometrical versions (Gr), (Fl) of the systems (Λ), (Λ_σ). We introduced a useful technical tool, namely the Bruhat decomposition of the algebraic Grassmannian $\mathrm{Gr}^{(n),\mathrm{alg}}$ (for maps $S^2 \to U_n$), or the algebraic flag manifold $\mathrm{Fl}^{(n),\mathrm{alg}}_{n-k,k}$ (for maps $S^2 \to \mathrm{Gr}_k(\mathbf{C}^n)$). By way of illustration, we used this to give a new proof of the classification of harmonic maps $S^2 \to \mathbf{C}P^n$. We also obtained estimates of the uniton number, and explicit Weierstrass formulae, this way.

Harmonic maps of finite type: The other main class of solutions $F(z,\lambda)$ that we consider is the class consisting of harmonic maps of finite type. In this case F is no longer assumed to have a finite Fourier series, but it is given in terms of a finite series $V(\lambda) = \sum_{i=-d}^{d} V_i \lambda^i$ by a factorization formula

$$F(z,\lambda) = [\exp z\lambda^{d-1}V(\lambda)]_1.$$

Here, each V_i is independent of z (and λ). The corresponding harmonic map $\phi = [\psi]$ is said to be of type d. Harmonic maps of finite type were introduced in Pinkall and Sterling [1989]; Ferus *et al.* [1992]; Burstall *et al.* [1993].

The most significant feature of such maps is that they can be constructed by combining two independent solutions of Lax equations, each of which is similar to the 1DTL (Chapters 23-24). Each Lax equation is defined on a finite dimensional subspace of a loop algebra, and is therefore amenable to explicit computation.

Finally, in Chapter 25, we sketched the theorem of Pinkall and Sterling [1989] that any harmonic map from a torus to S^2 is either of finite uniton number or of finite type. As in the case of harmonic maps of finite uniton number (see the bibliographical comments for Chapter 17), compactness of the domain is an essential ingredient.

It can be seen from this that loop groups provide a powerful tool for the study of harmonic maps. In fact, all known results on the classification problem may be accounted for by loop theoretic methods; and even in those cases where more traditional differential geometric methods are applicable, loop groups provide a unifying approach.

IV. Further developments.

Despite the successes of loop group techniques, the classification problem for harmonic maps from compact Riemann surfaces into compact Lie groups or symmetric spaces is far from solved. The case of a Riemann surface of genus 0 (i.e., the sphere) is quite well understood, there are results for special target spaces in the case of genus 1, but there are no classification results at all[9] for surfaces of genus greater than 1.

It seems likely that the theory of integrable systems will contribute to further progress. In this section we indicate some current directions of

---

[9]There are several non-trivial examples; see the introduction to Burstall *et al.* [1993].

research which may justify this prediction.

The solutions of all differential equations considered in this book have a striking common feature: They are all of the form

$$X = [\exp X_0]_1$$

where X_0 represents initial data of some kind, and where the suffix 1 means that a factorization is to be performed. (Similar formulae are known for other important integrable systems as well.) There is some evidence that this phenomenon will extend to harmonic maps of Riemann surfaces of arbitrary genus. Namely, it is shown in Dorfmeister *et al.* [in press] that a solution[10] F of (Λ_σ) corresponding to any such harmonic map must have the form

$$F = [H]_1$$

where H is a meromorphic (matrix-valued) function. Understanding the precise nature of H in this generalized "Weierstrass formula" seems to be an interesting problem.

A related matter is the idea of dressing transformations for integrable systems. We have so far discussed dressing transformations for the 2DTL and the harmonic map equation without making any concrete applications (except for the very special dressing transformations used in Chapter 20). However, it is obvious that these dressing transformations are relevant to the study of the *space* of solutions. In Guest and Ohnita [1993]; Furuta *et al.* [1994] some applications are given for spaces of harmonic maps of finite uniton number. In Dorfmeister and Wu [1993]; Wu [1993]; Burstall and Pedit [1994], dressing orbits of harmonic maps of finite type are studied. The space of harmonic maps seems to be a rather complicated object, and only partial results are known at this time (even in the case of genus 0).

There is a well known method in the theory of integrable systems, which can be used to study harmonic maps of finite type. It is the method of "spectral curves". For a harmonic map of finite type with associated polynomial Killing field $X(z, \lambda)$, consider (for each fixed z) the algebraic curve

$$\det(X(z, \lambda) - \mu I) = 0$$

in the (λ, μ)-plane. Because X satisfies a Lax equation, this curve is in fact independent of z; it is called the spectral curve of the harmonic map. The "spectral parameter" λ thus acquires true geometrical significance, which opens up a new (algebraic geometrical) direction for the study of harmonic maps.

---

[10]In this case, $F : U \to (\Lambda G)_\sigma$, where U is the universal cover of the Riemann surface.

For general information on this method, and the concept of "algebraically completely integrable systems", we refer to Perelomov [1990]; the article of Dubrovin, Krichever and Novikov in Arnold and Novikov [1990]; Arnold and Novikov [1994]; Mulase [1994]. The theory has been applied to harmonic maps in Pinkall and Sterling [1989]; Hitchin [1990]; Bobenko [1991]; Ferus *et al.* [1992]; McIntosh [1995], although this area is still very much under development.

Using this method, the "factorization of exponentials" formulae discussed earlier can be expressed much more explicitly in terms of θ-functions. In addition, there is some hope that harmonic maps of finite type can be completely classified in terms of "spectral data" (McIntosh [1995]).

It remains to mention briefly one other direction. There is a close connection between soliton theory and harmonic maps, if one replaces the Riemannian manifold $\mathbf{C} = \mathbf{R}^2$ by the Lorentzian manifold $\mathbf{R}^{1,1}$ (Pohlmeyer [1976]; Beggs [1990]; Gu and Hu [1993]). It is shown in Terng [in press] how such harmonic maps can be accommodated in the Adler-Kostant-Symes framework, by using a loop algebra of a Kac-Moody Lie algebra. This should strengthen even further the connection between harmonic maps, loop groups, and integrable systems.

References

J.F. Adams, *Lectures on Lie Groups*, Benjamin, 1969.

M. Adler and P. van Moerbeke, *Completely integrable systems, Euclidean Lie algebras, and curves*, Adv. Math. **38** (1980a), 267–317.

M. Adler and P. van Moerbeke, *Linearization of Hamiltonian systems, Jacobi varieties and representation theory*, Adv. Math. **38** (1980b), 318–379.-

V. Arnold, *Mathematical Methods of Classical Mechanics*, Graduate Texts in Math. 60, Springer, 1978.

V.I. Arnold and S.P. Novikov (eds.), *Dynamical Systems IV*, Encyclopaedia of Mathematical Sciences 16, Springer, 1990.

V.I. Arnold and S.P. Novikov (eds.), *Dynamical Systems VII*, Encyclopaedia of Mathematical Sciences 16, Springer, 1994.

G. Arsenault and Y. Saint-Aubin, *The hidden symmetry of $U(n)$ principal σ models revisited : I. Explicit expressions for the generators*, Nonlinearity **2** (1989a), 571–591.

G. Arsenault and Y. Saint-Aubin, *The hidden symmetry of $U(n)$ principal σ models revisited : II. The algebraic structure*, Nonlinearity **2** (1989b), 593–607.

M.F. Atiyah (ed.), *Representation Theory of Lie Groups*, London Math. Soc. Lecture Notes 34, Cambridge Univ. Press, 1979.

M.F. Atiyah, *Convexity and commuting Hamiltonians*, Bull. London Math. Soc. **16** (1982), 1–15.

E.J. Beggs, *Solitons in the chiral equation*, Comm. Math. Phys. **128** (1990), 131–139.

M. Berger and B. Gostiaux, *Differential Geometry: Manifolds, Curves, and Surfaces*, Graduate Texts in Math. 115, Springer, 1988.

M.J. Bergvelt and M.A. Guest, *Actions of loop groups on harmonic maps*, Trans. Amer. Math. Soc. **326** (1991), 861–886.

M. Black, *Harmonic Maps into Homogeneous Spaces*, Research Notes in Math. 255, Pitman, 1991.

A.M. Bloch, R.W. Brockett, and T.S. Ratiu, *Completely integrable gradient flows*, Comm. Math. Phys. **147** (1992), 57–74.

A.I. Bobenko, *All constant mean curvature tori in $\mathbf{R}^3$, S^3, H^3 in terms of thetafunctions*, Math. Ann. **290** (1991), 209–245.

J. Bolton, F. Pedit, and L. Woodward, *Minimal surfaces and the affine Toda field model*, J. reine angew. Math. **459** (1995), 119–150.

J. Bolton and L. Woodward, *Congruence theorems for harmonic maps from a Riemann surface into $\mathbf{C}P^n$ and S^n*, J. London Math. Soc. **45** (1992), 363–376.

R. Bott, *An application of the Morse theory to the topology of Lie groups*, Bull. Soc. Math. France **84** (1956), 251–281.

T. Bröcker and T. tom Dieck, *Representations of Compact Lie Groups*, Graduate Texts in Math. 98, Springer, 1985.

R.L. Bryant, *Lie groups and twistor spaces*, Duke Math. J. **52** (1985), 223–261.

F.E. Burstall, *Harmonic tori in spheres and complex projective spaces*, J. reine angew. Math. **469** (1995), 149–177.

F.E. Burstall, D. Ferus, F. Pedit, and U. Pinkall, *Harmonic tori in symmetric spaces and commuting Hamiltonian systems on loop algebras*, Ann. of Math. **138** (1993), 173–212.

F.E. Burstall and M.A. Guest, *Harmonic two-spheres in compact symmetric spaces, revisited*, preprint, available from http://xxx.lanl.gov/abs/dg-ga/9606002.

F.E. Burstall and F. Pedit, *Harmonic maps via Adler-Kostant-Symes theory*, Harmonic Maps and Integrable Systems (A. Fordy and J.C. Wood, eds.), Vieweg, 1994, pp. 221–272.

F.E. Burstall and F. Pedit, *Dressing orbits of harmonic maps*, Duke Math. J. **80** (1995), 353–382.

F.E. Burstall and J.H. Rawnsley, *Twistor Theory for Riemannian Symmetric Spaces*, Lecture Notes in Math. 1424, Springer, 1990.

F.E. Burstall and J.C. Wood, *The construction of harmonic maps into complex Grassmannians*, J. Differential Geom. **23** (1986), 255–297.

R. Carter, G. Segal, and I. MacDonald, *Lectures on Lie Groups and Lie Algebras*, LMS Student Texts 32, Cambridge Univ. Press, 1995.

E. Date, M. Kashiwara, and T. Miwa, *Transformation groups for soliton equations II*, Proc. Japan Acad. Ser. A Math. Sci. **57** (1981), 387–392.

E. Date, M. Jimbo, M. Kashiwara, and T. Miwa, *Transformation groups for soliton equations III*, J. Phys. Soc. Japan **50** (1981), 3806–3812.

E. Date, M. Jimbo, M. Kashiwara, and T. Miwa, *Transformation groups for soliton equations IV*, Phys. D **4** (1982), 343–365.

E. Date, M. Jimbo, M. Kashiwara, and T. Miwa, *Transformation groups for soliton equations V*, Publ. Res. Inst. Math. Sci. **18** (1982), 1111–1119.

E. Date, M. Jimbo, M. Kashiwara, and T. Miwa, *Transformation groups for soliton equations VI*, J. Phys. Soc. Japan **50** (1981), 3813–3818.

E. Date, M. Jimbo, M. Kashiwara, and T. Miwa, *Transformation groups for soliton equations VII*, Publ. Res. Inst. Math. Sci. **18** (1982), 1077–1110.

P. Deift, T. Nanda, and C. Tomei, *Differential equations for the symmetric eigenvalue problem*, SIAM J. Numer. Anal. **20** (1983), 1–22.

A. Doliwa and A. Sym, *Nonlinear σ-models on spheres and Toda systems*, Phys. Lett. **A185** (1994), 453–460.

J. Dorfmeister, F. Pedit, and H. Wu, *Weierstrass type representation of harmonic maps into symmetric spaces*, in press, Comm. Anal. Geom.

J. Dorfmeister and H. Wu, *Constant mean curvature surfaces and loop groups*, J. reine angew. Math. **440** (1993), 43–76.

J. Eells and L. Lemaire, *A report on harmonic maps*, Bull. London Math. Soc. **10** (1978), 1–68.

J. Eells and L. Lemaire, *Selected Topics in Harmonic Maps*, CBMS Regional Conference Series 50, Amer. Math. Soc., 1983.

J. Eells and L. Lemaire, *Another report on harmonic maps*, Bull. London Math. Soc. **20** (1988), 385–524.

J. Eells and J.C. Wood, *Harmonic maps from surfaces into complex projective spaces*, Adv. Math. **49** (1983), 217–263.

L.D. Faddeev and L.A. Takhtajan, *Hamiltonian Methods in the Theory of Solitons*, Springer, 1987.

D. Ferus, F. Pedit, U. Pinkall, and I. Sterling, *Minimal tori in S^4*, J. reine angew. Math. **429** (1992), 1–47.

A.T. Fomenko and V.V. Trofimov, *Integrable Systems on Lie Algebras and Symmetric Spaces*, Advanced Studies in Contemporary Mathematics 2, Gordon and Breach, 1988.

A. Fordy and J.C. Wood (eds.), *Harmonic Maps and Integrable Systems*, Vieweg, 1994.

T. Frankel, *Fixed points and torsion on Kähler manifolds*, Ann. of Math. **70** (1959), 1–8.

D.S. Freed, *The geometry of loop groups*, J. Differential Geom. **28** (1988), 223–276.

K. Fujii, *Nonlinear Grassmann σ-models, Toda equations, and self-dual Einstein equations: supplements to previous papers*, Lett. Math. Phys. **27** (1993), 117–122.

M. Furuta, M.A. Guest, M. Kotani, and Y. Ohnita, *On the fundamental group of the space of harmonic 2-spheres in the n-sphere*, Math. Zeit. **215** (1994), 503–518.

R. Goodman and N.R. Wallach, *Structure and unitary cocycle representations of loop groups and the group of diffeomorphisms of the circle*, J. reine angew. Math. **347** (1984a), 69–133.

R. Goodman and N.R. Wallach, *Classical and quantum mechanical systems of Toda-Lattice type*, Comm. Math. Phys. **94** (1984b), 177–217.

C.-H. Gu (ed.), *Soliton Theory and its Applications*, Springer, 1995.

C.-H. Gu and H.-S. Hu, *The soliton behavior of the principal chiral fields*, Int. J. Mod. Phys. A (Proc. Supp.) **3A** (1993), 501–510.

C.-H. Gu and H.-S. Hu, *Explicit construction of harmonic maps from $\mathbf{R}^2$ to $U(n)$*, Chinese Ann. Math. Ser. B **16** (1995), 139–152.

M.A. Guest and Y. Ohnita, *Group actions and deformations for harmonic maps*, J. Math. Soc. Japan. **45** (1993), 671–704.

M.A. Guest and Y. Ohnita, *Loop group actions on harmonic maps and their applications*, Harmonic Maps and Integrable Systems (A. Fordy and J.C. Wood, eds.), Vieweg, 1994, pp. 273–292.

V. Guillemin and S. Sternberg, *Symplectic Techniques in Physics*, Cambridge Univ. Press, 1984.

S. Helgason, *Differential Geometry, Lie Groups, and Symmetric Spaces*, Academic Press, 1978.

N.J. Hitchin, *Harmonic maps from a 2-torus to the 3-sphere*, J. Differential Geom. **31** (1990), 627–710.

G.R. Jensen and R. Liao, *Families of flat minimal tori in $\mathbf{C}P^n$*, J. Differential Geom. **42** (1995), 113–132.

190   References

V.G. Kac, *Infinite Dimensional Lie Algebras*, Cambridge Univ. Press, 1990.

M. Kashiwara and T. Miwa, *Transformation groups for soliton equations I*, Proc. Japan Acad. Ser. A Math. Sci. **57** (1981), 342–347.

B. Kostant, *The solution to a generalized Toda lattice and representation theory*, Adv. Math. **34** (1979), 195–338.

J.-H. Lu and A. Weinstein, *Poisson Lie groups, dressing transformations, and Bruhat decompositions*, J. Differential Geom. **31** (1990), 501–526.

I. McIntosh, *Global solutions of the elliptic 2D periodic Toda lattice*, Nonlinearity **7** (1994a), 85–108 (with correction in **8** (1995), 629–630).

I. McIntosh, *Infinite dimensional Lie groups and the two-dimensional Toda lattice*, Harmonic Maps and Integrable Systems (A. Fordy and J.C. Wood, eds.), Vieweg, 1994b, pp. 205–220.

I. McIntosh, *A construction of all non-isotropic harmonic tori in complex projective space*, Internat. J. Math. **6** (1995), 831–879 (with correction in Internat. J. Math., in press).

J. Milnor, *Morse Theory*, Ann. of Math. Stud. 51, Princeton Univ. Press, 1963.

J. Milnor, *Remarks on infinite dimensional Lie groups*, Relativity, Groups and Topology II (B. S. de Witt and R. Stora, eds.), North-Holland, 1984, pp. 1007–1057.

J.W. Milnor and J.D. Stasheff, *Characteristic Classes*, Ann. of Math. Stud. 76, Princeton Univ. Press, 1974.

R. Miyaoka, *The family of isometric superconformal harmonic maps and the affine Toda equations*, in press, J. reine angew. Math.

J. Moser, *Finitely many mass points on the line under the influence of an exponential potential - an integrable system*, Dynamical Systems, Theory and Applications (J. Moser, ed.), Lecture Notes in Physics 38, Springer, 1975, pp. 467–497.

M. Mulase, *Algebraic theory of the KP equations*, Perspectives in Mathematical Physics, Conf. Proc. Lecture Notes Math. Phys. III, International Press, 1994, pp. 151–217.

A.C. Newell, *Solitons in Mathematics and Physics*, CBMS-NSF Regional Conference Series in Applied Math. *48*, Soc. for Industrial and Applied Math., 1985.

S. Novikov, S.V. Manakov, L.P. Pitaevskii, and V.E. Zakharov, *Theory of Solitons*, Consultants Bureau, 1984.

G.D. Parker, *Morse theory on Kähler homogeneous spaces*, Proc. Amer. Math. Soc. **34** (1972), 586–590.

A.M. Perelomov, *Integrable Systems of Classical Mechanics and Lie Algebras*, Birkhäuser, 1990.

B. Piette and W.J. Zakrzewski, *General solutions of the U(3) and U(4) chiral σ models in two dimensions*, Nuclear Phys. B **300** (1988), 207–222.

U. Pinkall and I. Sterling, *On the classification of constant mean curvature tori*, Ann. of Math. **130** (1989), 407–451.

K. Pohlmeyer, *Integrable Hamiltonian systems and interactions through quadratic constraints*, Comm. Math. Phys. **46** (1976), 207–221.

M. Postnikov, *Lie Groups and Lie Algebras*, Mir, 1986.

A.N. Pressley, *The energy flow on the loop space of a compact Lie group*, J. London Math. Soc. **26** (1982), 557–566.

A.N. Pressley and G.B. Segal, *Loop Groups*, Oxford Univ. Press, 1986.

J. Ramanathan, *Harmonic maps from S^2 to $G(2,4)$*, J. Differential Geom. **19** (1984), 207–219.

J. Rawnsley, *Harmonic 2-spheres*, Quantum Theory and Geometry (M. Cahen and M. Flato, eds.), Kluwer, 1988, pp. 175–189.

H. Sakagawa, *The uniton number of harmonic 2-spheres in $Gr_2(\mathbf{C}^4)$*, in press, J. Math. Soc. Japan.

M. Sato, *Soliton equations as dynamical systems on infinite dimensional Grassmannian manifolds*, RIMS Kokyuroku **439** (1981), 30–46.

G. Segal, *An introduction to the paper "Schubert cells and cohomology of the spaces G/P"*, Representation Theory, London Math. Soc. Lecture Notes 69 (I.M. Gelfand et al., eds.), Cambridge Univ. Press, 1982, pp. 111–114.

G.B. Segal, *Loop groups and harmonic maps*, Advances in Homotopy Theory, London Math. Soc. Lecture Notes 139, Cambridge Univ. Press, 1989, pp. 153–164.

G. Segal and G. Wilson, *Loop groups and equations of KdV type*, Publ. Math. IHES **61** (1985), 5–65.

M.A. Semenov-Tian-Shansky, *Group-theoretical aspects of completely integrable systems*, Twistor Geometry and Non-linear Systems (H.D. Doebner and T.D. Palev, eds.), Lecture Notes in Math. 970, Springer, 1982, pp. 173–185.

M. Spivak, *A Comprehensive Introduction to Differential Geometry, Vol. 1*, Publish or Perish, 1979.

S. Sternberg, *Lectures on Differential Geometry*, Prentice-Hall, 1964.

G. Strang, *Linear Algebra and its Applications*, Academic Press, 1976.

C.-L. Terng, *Soliton equations and differential geometry*, in press, J. Differential Geom.

K.K. Uhlenbeck, *Harmonic maps into Lie groups (Classical solutions of the chiral model)*, J. Differential Geom. **30** (1989), 1–50.

H. Urakawa, *Calculus of Variations and Harmonic Maps*, Translations of Math. Monographs 132, Amer. Math. Soc., 1993.

G. Valli, *On the energy spectrum of harmonic 2-spheres in unitary groups*, Topology **27** (1988), 129–136.

V.S. Varadarajan, *Lie Groups, Lie Algebras, and Their Representations*, Graduate Texts in Math. 102, Springer, 1984.

R.S. Ward, *Classical solutions of the chiral model, unitons, and holomorphic vector bundles*, Comm. Math. Phys. **128** (1990), 319–332.

F.W. Warner, *Foundations of Differentiable Manifolds and Lie Groups*, Graduate Texts in Math. 94, Springer, 1984.

H.C. Wente, *Counterexample to a conjecture of H. Hopf*, Pacific J. Math. **121** (1986), 193–243.

G. Wilson, *The modified Lax and two-dimensional Toda lattice equations associated with simple Lie algebras*, Ergodic Theory Dynamical Systems **1** (1981), 361–380.

G. Wilson, *Infinite dimensional Lie groups and algebraic geometry in soliton theory*, Phil. Trans. R. Soc. Lond. **A315** (1985), 393–404.

J.G. Wolfson, *On minimal surfaces in Kähler manifolds of constant holomorphic sectional curvature*, Trans. Amer. Math. Soc. **290** (1985), 627–646.

J.C. Wood, *Explicit construction and parametrization of harmonic two-spheres in the unitary group*, Proc. London Math. Soc. **59** (1989), 608–624.

H. Wu, *Banach manifolds of minimal surfaces in the 4-sphere*, Differential Geometry, Proc. Symp. Pure Math. 54 (R.E. Greene and S.T. Yau, eds.), Amer. Math. Soc., 1993, pp. 513–539.

V.E. Zakharov and A.V. Mikhailov, *Relativistically invariant two-dimensional models of field theory which are integrable by means of the inverse scattering problem method*, Soviet Phys. JETP **47** (1978), 1017–1027.

V.E. Zakharov and A.B. Shabat, *Integration of non-linear equations of mathematical physics by the inverse scattering method II*, Funct. Anal. Appl. **13** (1979), 166–174.

Index

Toda lattice, one-dimensional, ix, 26, 36, 39
Toda lattice, two-dimensional, 55, 62, 77, 94, 139, 172
twistor construction, 145

uniton, 108
uniton number, 109, 133

vector field, 2

Weierstrass formula, 146

zero-curvature equation, 54, 158